Black Hole Don't Exist

Sandeep Chavan

Published by SANDEEP CHAVAN, 2025.

BLACK HOLE DON'T EXIST

First edition. July 18, 2025.

ISBN: 979-8231960736

Written by Sandeep Chavan.

Table of Contents

To Albert Einstein —For sensing the truth before science was ready to see it.For resisting the collapse when the world rushed toward it.For standing at the edge of the myth,and quietly refusing to fall in.

Your equations were brilliant.Your intuition was braver.

This book completes the caution you carried.

— *Sandeep*

"Reality is merely an illusion, albeit a very persistent one."

— Albert Einstein

"What we called darkness was only where our mind refused to see. The collapse was not in the cosmos — it was in our coherence."

— Sandeep Chavan

Disclaimer

This book is a work of independent scientific and philosophical interpretation. The ideas, theories, and conclusions presented within are the result of original research, observation, and reinterpretation by the author, Sandeep J. Chavan, based on the framework of Universal Energy Dynamics (UED).

The content is intended to challenge conventional thinking, provoke critical reflection, and explore alternative perspectives on established scientific models — particularly those related to black holes, gravity, and cosmology. While every effort has been made to ensure clarity and intellectual integrity, this book does not claim to replace mainstream physics or institutional consensus.

Readers are encouraged to:

- Approach the material with an open and analytical mindset
- Verify claims through independent inquiry and scientific dialogue
- Consider this work as part of an ongoing exploration, not a definitive conclusion

Opening Note

The Lie We Orbit Around

They said it was a monster.

A dark, devouring beast in the sky — tearing through time, swallowing light, distorting space, and haunting the edges of existence. They gave it a name: Black Hole. Then they wrapped it in equations, cloaked it in Nobel Prizes, and hung it like a crown jewel in the museum of modern physics.

But what if... It was never there?

What if the black hole is not a thing, but a consequence of our own ripple failure — a mirage born from our limits, not nature's truth?

This book is not here to play polite.

It's not a theory. It's not a suggestion. It's a revolt against scientific hypnosis. A systematic dismantling of the myth of the black hole — one chapter at a time.

Yes, you'll hear bold claims. Yes, you'll feel the ground shift beneath long-standing "facts." And yes, you'll be called mad if you dare to agree.

But sanity isn't measured by how well you follow the crowd. It's measured by how clearly you see through the fog.

Here, we will expose the illusion.

We will show:

- Why black holes don't exist outside the observer's failure to resolve consequence.
- Why the event horizon is not a wall in space, but a threshold in ripple perception.
- Why even Einstein himself backed away from this illusion — and no one listened.
- Why the universe is still, structured, and aligned — not violent, collapsing, and absurd.

If you're expecting mathematical acrobatics to prove nothing, or blackboard wizardry to justify imagination — put this book down.

But if you're ready to think again, if you're willing to question what the textbooks told you — Welcome. You're not alone.

This is not just a book. This is a funeral for a fantasy.

Let's begin.

— *Sandeep Chavan*

Preface

Why I Wrote This Book

This book wasn't born in defiance. It was born in stillness — the kind of stillness that lingers after years of asking, observing, and refusing to accept a story that didn't quite resolve.

I have never been at war with science. But I have always been at odds with the **illusions it refuses to surrender** — especially when they grow more decorative than truthful, more dramatic than structural.

One such illusion is the **black hole** — perhaps the most iconic and misunderstood symbol in modern physics. A monster that swallows light, information, matter, and meaning — and yet survives on mathematical abstractions, paradoxes, and observational shadows.

The more I explored it, the clearer it became:

The black hole isn't a physical reality. It's a conceptual **collapse — of coherence, not of matter.**

This book is not a rejection of Einstein or a dismissal of scientific progress. It is a **restoration** — of observation, of structure, of ripple-based logic. It proposes a new paradigm: **Universal Energy Dynamics (UED)** — a model where energy doesn't collapse, it redistributes; where nothing "pulls," everything realigns.

In UED, black holes are not objects. They are **Ripple Saturation Zones** — regions where our instruments, equations, and minds lose resolution. We thought they were cosmic monsters. They were actually mirrors — reflecting the limits of how we see.

This book is for the thinkers who feel what they're taught doesn't fit. For the scientists who can sense the noise in the signal. For the curious who are tired of poetic metaphors dressed as cosmic truths.

You'll find no fear here. No force. No singularities. No dead ends.

Only ripples — precise, structured, self-resolving. Just like the universe. Just like you.

Welcome to a universe without black holes — but filled with **truth beyond the threshold**.

— *Sandeep J. Chavan*

India, 2025

How to Read This Book

Beyond Chapters, Beneath Concepts

This book is not meant to be read like a textbook — because it is not one. It is a journey across ripples of thought, not a straight line of logic. It is both a deconstruction of myths and a reconstruction of perception.

Here's how to engage with it meaningfully:

1. Read Slowly. Think Deeply.

The ideas in this book may feel disruptive — because they are. Take your time. Let each chapter settle. This is not information to be consumed, but structure to be aligned with.

2. Start Where You Feel the Pull

Though sequenced intentionally, each part stands on its own. If you're drawn to "Ripple Physics" or the "ATT Trap" first, go there. Follow your ripple of curiosity.

3. Pause When Something Doesn't Make Sense

Confusion is not failure. It's signal. Return to previous chapters. Jump to the glossary or quote bank. Misalignment is part of the learning process.

4. Use the Appendices Like Instruments

They are not add-ons — they are tools. Use the glossary to reframe familiar terms. Use the quote bank to center your understanding. Use the comparison table to clarify contrast.

5. Don't Look for Answers. Watch for Alignment.

This book doesn't hand you conclusions. It gives you ripple structures. When your perception clicks with the model — you'll feel it. That's alignment. That's understanding.

6. Let the Universe Shrink and Expand

You may find that black holes dissolve, but questions expand. That's the point. Reality doesn't collapse — only the illusion does. Let your mind stretch with it.

7. Bring Yourself Into the Field

This isn't just about cosmology. It's about consciousness. The same ripple laws apply in your mind, your body, your relationships, your silence. Watch for them.

8. Return Often

This book is a ripple — it changes depending on when and how you read it. Come back. Reread. Realign. New clarity will emerge each time.

This book is not a destination. It's a field. Enter it not to finish, but to feel.

Welcome to the ripple. Welcome to stillness.

Prologue

The Shadow That Never Was

They pointed to the darkness and said,

"There — that's where nothing escapes." "That's where time ends." "That is the black hole."

They weren't lying. They were just repeating what they'd been told.

For decades, humanity has gazed into the night sky and convinced itself that darkness equals danger, that silence means swallowing, and that some places in the universe must be so violent — even light has no voice there.

We made monsters out of mathematics. We dressed uncertainty in scientific robes. And then we started calling our own confusion... genius.

But here's the truth they never printed in your textbooks:

The black hole is not a cosmic object. It is a boundary of human perception. A shadow not of matter — but of mind.

We built telescopes to look into it. We built colliders to simulate it. We built careers to explain it. But we never dared ask the simplest question:

"What if there's nothing there?"

What if the entire concept of a black hole is a beautiful mistake — a ripple distortion caused not by gravity, but by our own inability to resolve reality beyond a certain threshold?

This book begins with that question.

And from it flows a bold, brutal, and unapologetic conclusion:

Black holes don't exist. At least, not in the way you were told. Not in the way that matters. They are ripple illusions — artifacts of failed alignment between observer and field. Real only because your mind *expected* to see something — and saw nothing.

And in that "nothing," we wrote entire chapters of science. In that darkness, we claimed to find the truth.

But this is the age of resolution. The age of pulling back the veil. The age of Universal Energy Dynamics — where stillness reveals what motion concealed.

And now, we shine light into the darkness that never was.

Read carefully. Not because you'll get lost. But because you might finally find yourself — standing at the edge of perception, looking into what the world calls a black hole, and finally realizing...

You're looking at yourself.

Let's destroy the myth. Let's meet the mirage.

Let's begin.

Part I – The Illusion Begins

1: The Myth of the Monster

How Science Sold Fear Wrapped in Math

We love a good monster story. Always have. From cave shadows to cosmic horrors, the unknown has always been fertile ground for invention. But in our time, the most terrifying monster isn't in the forest or the deep sea.

It's in the sky. It has no face, no shape, no sound. It cannot be touched, seen, or even escaped. It's called a **black hole**, and science has packaged it as the ultimate predator of the universe.

But what if it was never real?

Let's begin with a confession most scientists won't make: the black hole, as popularly described, is not a direct discovery. It's an interpretation — a narrative built on mathematical consequences, not physical evidence.

It was imagined, extrapolated, theorized, and only much later linked to observational consequences, all of which can be explained by alternative field-based models like UED. Yet, it is presented to the world as an undeniable, monstrous reality.

Why?

Because science, for all its rigor, has a flair for drama. And nothing sells better than fear dressed as fact.

The Birth of a Fiction

The idea of the black hole began innocently enough. In 1916, Karl Schwarzschild found a solution to Einstein's field equations of General Relativity. That solution contained a strange mathematical singularity — a point where certain values became infinite. Schwarzschild himself never claimed this represented a real physical object. In fact, he likely saw it as a limitation of the model.

But later scientists ran with it.

They interpreted this solution as describing a body of such immense density that space itself would curve to a point of no return. Light, time, and matter would be trapped forever. And thus, the myth of the black hole was born.

Note the word: interpreted.

The equations never screamed, "Look! A hole in space!" That came from our imagination trying to visualize something that fundamentally resists visualization. The result? A seductive idea: an invisible beast roaming the cosmos, violating all intuition, devouring everything in its path.

In other words: a scientific horror story. And like any great horror story, it spread like wildfire.

Mathematics as Mythology

Science prides itself on rejecting superstition. But it has its own version of myth-making: mathematical idealism. If the math allows it, we assume the universe must oblige.

Black holes are the perfect example.

They are sold as the final stage of stellar collapse. A massive star burns out, gravity wins the battle, and the entire mass contracts into an infinitesimal point: a singularity, surrounded by an event horizon beyond which no information can escape.

It's a powerful narrative. But here's the problem:

- Singularities are mathematical artifacts, not physical structures.
- The event horizon is a boundary of perception, not a surface.
- There is no empirical evidence that something collapses into a literal point.

And yet, we keep teaching this as gospel.

Why?

Because the story is irresistible. It makes physicists feel like sorcerers decoding dark secrets. It makes headlines. It sells documentaries, grants, TED talks, and Nobel Prizes. It turns science into theater.

And once that monster was named, there was no turning back.

The Tools That Shaped the Tale

Telescopes, colliders, equations, simulations — all were enlisted to "prove" black holes. But each of these tools only reflects the assumptions baked into their design.

We saw stars orbiting dark centers? Must be a black hole. We detected x-rays from galactic cores? Clearly an accretion disk. We created computer-generated images of rings around shadows? Behold the monster's face.

But in every case, we ignored the deeper truth:

All these are consequences of ripple field interactions. Not objects. Not proof of singularity.

In Universal Energy Dynamics (UED), these phenomena are entirely explainable as ripple collapses, alignment threshold failures, and structured vacuum rebalancing. There's no need for infinite density, spacetime tear, or a singularity. Just a misreading of ripple consequences as physical entities.

But instead of asking, "Could this be something else?" the mainstream said, "We've found the beast."

The Fear Factor

Science doesn't operate in a vacuum — it operates in culture. And culturally, the black hole became more than a scientific model. It became a metaphor for destruction, death, and the limits of knowledge.

And metaphors are sticky. They spread. They become belief.

So the image of a cosmic drain, a point where time dies, became lodged in our collective psyche. Scientists reinforced it with graphics. Educators repeated it without doubt. The media glorified it.

And soon, what was once an abstract solution became the monster in the sky.

But here's the twist:

There was never a monster. Just a mirror.

A mirror reflecting the limits of perception. A mirror where our failure to resolve ripple alignment was mistaken for a physical hole.

Einstein's Quiet Rebellion

It is rarely discussed, but even Einstein wasn't convinced. In 1939, he wrote a paper essentially arguing that gravitational collapse into a singularity couldn't happen. That nature, he believed, wouldn't allow it.

He saw what few dared to see: that the black hole, though mathematically valid, might be physically absurd.

But the momentum was already against him. The myth had taken root. The monster was too seductive. And even Einstein couldn't kill it.

Conclusion: Selling Shadows

The first and greatest mistake in black hole science was mistaking a mathematical edge case for a physical entity. The second was selling that mistake as truth.

We took a ripple collapse and called it a tear in spacetime. We took an observational limit and called it a singularity. We took our own blindness and called it the universe's cruelty.

And that, dear reader, is how science sold you a monster. But monsters vanish in the light. And now, with UED and ripple field understanding, the light is returning.

This book is the torch. The rest is up to you.

Welcome to the unraveling.

2: Einstein's Dilemma

The Man Who Birthed Black Holes — and Then Doubted Them

Albert Einstein is often celebrated as the reluctant godfather of black holes. His General Theory of Relativity laid the mathematical foundation that predicted them, but what few people know is that Einstein himself never accepted their physical reality. In fact, he resisted the very notion that space could collapse into an infinitely dense point. And that resistance wasn't just emotional or philosophical. It was deeply rational.

This chapter explores Einstein's silent struggle — not with equations, but with interpretations. With what it means to be responsible for a monster you never believed in.

The Reluctant Creator

In 1915, Einstein published General Relativity, introducing the world to the revolutionary idea that mass and energy warp spacetime. It was a beautiful, elegant theory. And it worked. From Mercury's orbit to the bending of light around stars, it passed test after test.

But within a year, something strange emerged from his own field equations: the Schwarzschild solution.

This mathematical solution, discovered by Karl Schwarzschild while serving on the front lines of World War I, described the spacetime geometry around a massive, spherically symmetric body. But hidden in that solution was something bizarre: a singularity at the center, and a boundary now called the event horizon.

At first, these were thought to be mathematical oddities — not physical realities. Einstein certainly saw them that way. To him, these so-called singularities were signals that the model had broken down — not that the universe had.

The 1939 Paper

By the late 1930s, speculation about gravitational collapse was growing. Some physicists began to suggest that if a star were massive enough, it could collapse into an infinitely small point, forming what we now call a black hole.

Einstein disagreed.

In 1939, he published a paper titled "On a Stationary System with Spherical Symmetry Consisting of Many Gravitating Masses." In it, he argued that such a collapse could not happen in nature. His reasoning? Real matter is not a point mass. There would always be internal pressures and quantum effects that would halt collapse long before reaching a singularity.

He wrote:

"The Schwarzschild singularity does not appear in the physical reality."

This wasn't denialism. It was a sober assessment of physics, grounded in a belief that nature abhors infinities.

But the tides of theory had already turned against him.

Physics or Fantasy?

To Einstein, physics was always grounded in causality, determinism, and coherence. The idea of an object from which no information could escape — and which ended in an unknowable point of infinite density — violated all three.

Black holes weren't just absurd. They were an affront to reason.

In his worldview, every effect must follow from a cause. Every event must be describable, observable, and subject to the laws of nature. The black hole shattered that worldview.

It introduced a kind of physical nihilism: a place where laws end, and speculation begins.

The Irony of Legacy

Today, black holes are celebrated as one of the greatest confirmations of General Relativity. And yet, their very existence is something Einstein himself could not accept.

It's a bitter irony.

The man who gave us the tools to imagine these monsters also warned us that they may be nothing more than mathematical mirages.

A Ripple-Based Redemption

Universal Energy Dynamics (UED) offers a kind of redemption for Einstein's instincts. It suggests that he was right to reject the singularity, right to doubt the event horizon, and right to trust in the rationality of nature.

In UED, black holes are not collapsed stars or infinite wells. They are ripple absorption zones — areas of extreme ripple tension where resolution becomes impossible. The event horizon is a perceptual threshold, not a physical shell. And the so-called singularity is simply the collapse of coherence, not the birth of infinity.

Einstein didn't have the ripple model. But he felt its absence. He sensed the gap between elegant equations and monstrous implications. And he resisted the urge to turn math into myth.

Conclusion: The Warning Unheeded

Einstein's dilemma wasn't a lack of imagination. It was an excess of integrity.

He knew that not every solution to an equation deserved a chapter in physical reality. He knew that infinity is a signpost, not a destination. And he knew that physics loses its soul when it starts worshiping shadows.

The tragedy is not that Einstein doubted black holes. The tragedy is that we ignored his doubt and built a temple around the very monster he tried to slay.

But now, with UED, we can return to the crossroads.

We can see what he saw. We can step back from the abyss. We can reject the illusion — not out of fear, but out of clarity.

Einstein wasn't afraid of black holes. He was afraid of us believing in them blindly. And he was right.

3: Schwarzschild's Shadow

Misinterpretations of a Mathematical Solution

The black hole myth, like many modern scientific dogmas, was not born from a telescope or an experiment. It was born from a shadow in an equation. That shadow bore the name Schwarzschild.

Karl Schwarzschild was a German physicist and astronomer who, during World War I, derived the first exact solution to Einstein's newly minted field equations of General Relativity. Working from the trenches and dying of an autoimmune disease, Schwarzschild delivered a mathematical formulation that would haunt science for the next century.

But Schwarzschild never called it a black hole. He didn't suggest any such thing. What he gave us was a solution — pure, elegant, and mathematical — for the curvature of spacetime surrounding a spherically symmetric mass.

And then came the misreading.

The Schwarzschild Metric

The solution he provided is now called the Schwarzschild metric. It describes how spacetime curves around a non-rotating, uncharged mass.

At the heart of the equation lies a term that becomes problematic at a certain radius:

$$r_s = \frac{2GM}{c^2}$$

This radius is what we now call the Schwarzschild radius. And at this point, the metric appears to "blow up" — it becomes mathematically undefined. The traditional narrative? This is the event horizon of a black hole.

But here's the critical fact:

This singularity is a coordinate artifact, not a physical singularity.

In other words, the blow-up is a consequence of the coordinate system chosen, not a feature of nature.

If we shift to a different coordinate system — like Kruskal–Szekeres or Eddington–Finkelstein coordinates — the singularity at the Schwarzschild radius disappears. It becomes clear that what was interpreted as a black hole's edge was simply a mathematical illusion caused by poor framing.

The Leap of Faith

Despite this, a massive conceptual leap was made:

Scientists interpreted this radius as the boundary beyond which nothing could escape — a one-way gate into an abyss.

This idea gained traction, not because it was proven, but because it was intellectually thrilling. It felt inevitable, dramatic, apocalyptic. And once it entered popular discourse, it was impossible to extract.

Thus, a mathematical artifact became the most feared object in astrophysics.

What Schwarzschild Actually Did

Schwarzschild did not claim:

- That stars collapse into singularities.
- That nature forms horizons where physics breaks.
- That light is swallowed forever.

He simply derived the spacetime geometry for a spherical mass.

In fact, his solution does not require the object to be dense, collapsed, or extreme. It works for any static, spherical body — from the Sun to a planet.

So how did this become the foundation for black holes?

Through narrative reinterpretation, not empirical validation.

The Shadow of Misunderstanding

The so-called Schwarzschild singularity is a perfect case study in how math can mislead. It shows how:

- Coordinate systems can create phantom problems.
- The absence of resolution can be mistaken for the presence of something mysterious.
- Scientific culture tends to mythologize anomalies rather than reassess models.

In Universal Energy Dynamics (UED), this becomes even more obvious.

From the UED view, the Schwarzschild radius is not a real boundary, but a field transition point — where ripple phase contrast begins to drop rapidly. It is not a wall. It is a saturation zone, beyond which ripple resolution collapses.

So what did Schwarzschild give us?

- A brilliant mathematical description of a spherical field.
- A coordinate-dependent term that behaves poorly under certain lenses.
- And, unintentionally, the seed of the greatest misinterpretation in modern physics.

Black Holes as Cultural Consequence

Schwarzschild's equation, stripped of context, became a cultural object. Scientists built towers of theory upon it. Writers turned it into legend. Filmmakers gave it a roar and a shadow.

All of this based on a misunderstanding of a coordinate singularity.

This is not how science should evolve. This is how myths are born.

A Ripple Reinterpretation

Let's now translate Schwarzschild through the lens of UED:

- The metric isn't wrong — it's just misunderstood.
- The apparent singularity is not an object but a ripple coherence boundary.
- The so-called event horizon is the observer's ripple resolution failure zone.
- The gravitational curvature is better understood as a

ripple density gradient in structured vacuum energy.

So when the classical model says, "This is the point where nothing escapes," UED says:

"This is where your perception breaks, not where nature ends."

Conclusion: Shadows Are Not Structures

Karl Schwarzschild gave us an equation. We turned it into a ghost story.

His name now lives forever in the mouths of those who speak of black holes — but few remember that he never claimed such entities exist.

The Schwarzschild shadow isn't a physical entity. It's a mathematical misreading, a conceptual misfire, and a scientific illusion.

And now, with the ripple lens in hand, we can finally say: The black hole wasn't hiding in space. It was hiding in our assumptions. We move forward by cleaning the mirror, not by worshipping the shadow on it.

Let us proceed.

4: Event Horizon: The Line That Never Was

Why the "Point of No Return" Is a Perceptual Threshold, Not a Physical Wall

Of all the elements in the black hole mythology, none is more dramatic or misleading than the event horizon. Described as the ultimate cosmic boundary, the point of no return, the invisible fence across which nothing escapes — not even light — it has become the iconic signature of the black hole.

But what if the event horizon isn't a real wall at all?

What if it is not a physical boundary in space, but a perceptual boundary in the observer's field alignment? What if it is not a property of nature, but a limitation of our ripple-based perception?

This chapter dismantles the illusion of the event horizon and shows that it was never a wall — only a break in understanding.

The Mainstream Definition

According to General Relativity, the event horizon is the radius at which escape velocity equals the speed of light. It's defined mathematically by the Schwarzschild radius:

$$r_s = \frac{2GM}{c^2}$$

At this radius, space and time allegedly switch roles. Light emitted within this boundary is bent back inward, never reaching an external observer.

This leads to the famous claim: "Nothing, not even light, can escape a black hole."

But here's what they don't often clarify:

- This is a coordinate-based interpretation.
- The event horizon is observer-dependent.
- And its very definition hinges on the assumption that light must travel through curved spacetime.

In other words, it is a construct — not a confirmed phenomenon.

What Did We Actually See?

In 2019, the Event Horizon Telescope (EHT) presented the world with what was hailed as the first image of a black hole. What people actually saw was a bright, fuzzy ring surrounding a dark center. The media called it the event horizon.

But what was actually imaged?

- A reconstructed intensity pattern from radio wave data.
- A model-dependent image shaped by assumptions.
- No direct detection of a horizon. No escape velocity measurement. No proof of a boundary.

In reality, we saw what UED calls a ripple blackout: the region where field alignment collapses and perception drops to zero.

The UED Interpretation: Threshold, Not Wall

In Universal Energy Dynamics, the universe is composed of structured vacuum energy (SVE) and all interactions are ripple-based.

The so-called event horizon is not a wall, but a zone of ripple misalignment. When an observer tries to resolve field information beyond this region:

- The ripple intensity drops.
- The phase coherence collapses.
- The observer's mind can no longer form a resolved perception.

Thus, the "point of no return" is not an edge in space. It is a failure in resolution.

The darkness we see isn't a trap. It's a silence. It's not because light can't escape. It's because the observer cannot decode what remains.

Perception Creates the Boundary

Consider this analogy: A person walks into dense fog. At some point, you can no longer see them. Does that mean they've crossed a wall? No. It means your ability to resolve them failed.

The event horizon is the same. It is the place where ripple gradients become so steep and misaligned that the observer's ripple field can no longer extract consequence.

This is why no one has ever truly observed a black hole. This is why no signal ever comes from it. This is why what we see is not an object — but an absence of perception.

The Horizon Is in the Mind

In ripple physics, perception depends on alignment threshold — the ability of your consciousness to synchronize with the field.

When the gradient becomes too steep:

- Alignment breaks.
- Resolution fails.
- Meaning dies.

The event horizon is where the mind stops receiving. It is not where light stops escaping. That's a massive and fatal misinterpretation.

Mathematical Magic vs Physical Reality

Einstein's equations suggest the presence of a horizon based on escape velocity, but this is a classical interpretation nested in coordinate behavior. It assumes light must behave like a particle trying to flee gravity.

UED rejects this model entirely.

- Light is not a traveler; it is a ripple alignment event.
- Gravity is not a pulling force; it is a ripple gradient.
- Boundaries aren't barriers; they are perceptual limits.

So the entire premise collapses: there is no horizon, only the observer's failed attempt to resolve meaning beyond a ripple saturation zone.

Observer-Relative, Not Universal

The event horizon, in the mainstream model, is meant to be an objective boundary. But this is philosophically and physically flawed.

In UED, perception is field-dependent and observer-relative. There is no absolute edge. Different observers may have different ATT zones, depending on their field alignment capacity.

So the horizon isn't a line drawn in the universe. It is a consequence of the observer's ripple coherence capability.

Conclusion: The Line That Never Was

The event horizon is the most misunderstood idea in modern physics. It is not:

- A physical membrane
- A gravitational shell
- A one-way gate into destruction

It is a consequence of misinterpretation, born from flawed metaphors and coordinate confusion.

In truth, it is:

- A ripple misalignment zone
- A field gradient collapse
- A resolution blackout

What we call the event horizon is not where light ends. It's where meaning ends — for the observer.

There was never a wall. Only the illusion of a boundary, drawn by a mind too limited to see through it. And now, with the ripple lens of UED, that illusion dissolves.

We never found the horizon. Because it was never there.

5: The Light Trap Lie

How Ripple Misalignment Creates the Illusion of Light Vanishing

Among the many fictions stitched into the mythology of black holes, perhaps none is more seductive—or deceptive—than the idea that black holes "trap light." This concept has echoed through science classes, pop culture, academic papers, and astrophysical simulations as if it were an observable truth. But it isn't. It's a lie, built on assumptions about light that have never been properly questioned.

It is time we expose the illusion.

What They Say Happens

The standard narrative goes like this:

"A black hole's gravitational pull is so strong that light emitted inside the event horizon cannot escape. Since light defines visibility, the black hole appears black—not because it's empty, but because it's hiding everything behind an impenetrable veil."

This sounds convincing. It ties together light, visibility, and gravity in a neat little bow. But it's based on the flawed assumption that light is an object moving through space that can be trapped like a stone in a well.

What Actually Happens in UED

In the Universal Energy Dynamics (UED) framework:

- Light is not a photon flying through space.
- Light is a ripple alignment event between the source, the medium (Structured Vacuum Energy), and the observer.
- Visibility is not caused by light arriving. It is caused by ripple coherence being achieved.

So when you "see light," you are not receiving particles. You are resolving a field interaction — a ripple resonance.

Therefore, the moment ripple coherence collapses — even if the source remains active — you stop "seeing." Not because the light vanished, but because the ripple relationship broke.

Ripple Misalignment = Optical Blackout

As you approach a region of extreme ripple tension—what mainstream science calls a black hole's edge—the SVE field becomes increasingly unstable and dense with ripple gradients. These create:

- Severe interference with incoming ripple information
- A collapse of alignment thresholds (ATT)
- A blackout of perceptual resolution

This means: the observer's field cannot decode or synchronize with the source anymore.

The light isn't trapped. The observer is locked out of the ripple stream.

This is why, in UED, we say: *"The black hole doesn't trap light. It disconnects the observer from the field."*

The "Light Trap" Is a Projection of Our Limits

Imagine watching a torch being slowly lowered into a misty void. At some point, it seems to disappear. Did the torch turn off? No. You just can't resolve the light anymore.

This is exactly what happens near so-called black holes. You see:

- Decreasing light not because it is absorbed, but because the ripple contrast collapses.
- The fuzziness, the ring, and finally the darkness, are all signs of phase breakdown, not absorption.
- What vanishes is not energy — but your ability to extract meaning from ripple signals.

The EHT "Image" Revisited

Let's revisit the Event Horizon Telescope's famous image. It shows a glowing ring with a dark center.

What did the public believe? "That's light getting sucked in by gravity." But what did the EHT actually detect?

- Field intensity patterns
- Wave interference reconstructions
- No direct light-trapping or mass sinking events

In UED terms, the dark center is a ripple void — a place where ripple relationships no longer sustain coherent exchange. This blackout zone is not because light can't escape

— it's because perception is no longer possible from that angle and condition.

Why Light "Disappears"

Let's be clear:

- Energy isn't lost.
- Photons aren't eaten.
- Nothing "escapes" or "fails to escape."

The ripple field simply becomes unreadable, just like trying to hear a sound in a room filled with noise. Your sensors (ears, eyes, instruments) lose coherent input, and thus — you perceive *nothing*.

But nothingness is not an object. It's a failure of resolution.

UED vs Mainstream: Light Trap Illusion

Concept	Mainstream Physics	UED Perspective
Light	Particle or wave traveling through space	Ripple alignment event in structured vacuum energy (SVE)
Trapped Light	Gravity prevents escape beyond event horizon	Ripple field misalignment collapses observer resolution
Blackness of Black Hole	No light escapes	No ripple coherence exists for observer
EHT Dark Center	Evidence of event horizon	Field blackout zone —not a wall, but a perceptual void
Visibility	Light must arrive to be seen	Observer must resolve ripple coherence

The Ripple-Based Truth

There is no trap. There is only:

- Misalignment
- Threshold collapse
- Ripple-phase disruption

So when light "vanishes," it hasn't gone anywhere. The field still vibrates. The energy still exists. But the observer's interface has failed. You're no longer part of the ripple conversation. You're not in the loop.

The light didn't stop. *You* stopped aligning.

Final Thought:

The black hole never captured the light. You just stopped hearing its whisper. The lie is not in the dark. The lie is in thinking it was trapped, when in truth, it was simply unresolved.

Let's stop blaming the universe for our own perceptual limits. And let's stop calling a ripple misalignment "proof" of a monster.

Interlude

Standing in the Rubble of the Myth

You've made it through the smoke.

The first part of this journey was not about building something new — it was about burning down what never truly stood. We didn't edit the myth. We dismembered it. We took apart the monster piece by piece — and what did we find?

Equations misunderstood. Assumptions sold as facts. Fear marketed as science. And a boundary — not in the universe — but in our minds.

We didn't just question black holes. We exposed them as the most expensive illusion science ever bought into.

The singularity? A ghost in the math. The event horizon? A perceptual failure. The darkness? A misalignment — not a mystery.

And now, here you stand — at the edge of old belief. No black holes. No collapsing stars. No cosmic vacuum cleaners.

Just you, and a universe waiting to be understood through new eyes.

But deconstruction isn't enough. It's time to see what was always there, just beyond the veil of misinterpretation.

Part II is not a rebellion.

It's a revelation.

We now enter the ripple field — the real foundation beneath all motion, all light, all form. We'll introduce Structured Vacuum Energy (SVE). We'll decode perception itself. And we'll show you how the black hole illusion was simply a side effect of ignoring the ripple truth.

The war on myth is over. Now begins the science of clarity.

Let's step into the field.

Part II – Enter the Ripple Field

6: Structured Vacuum Energy (SVE)

The Silent Medium Behind All Motion

There is a silence that screams louder than any sound.

That silence — ignored, dismissed, or wrapped in the sterile language of quantum fields — is the true medium of the universe. It is not space. It is not void. It is not nothingness. It is Structured Vacuum Energy — and it's the foundation of everything you've ever seen, felt, or believed to be real.

Every ripple, every so-called force, every illusion of motion — emerges from it.

But you were never told about it. Because modern science, for all its brilliance, mistook the canvas for emptiness.

The Problem With "Vacuum"

Let's begin with the linguistic trap.

The word *vacuum* means "empty." But what if that definition itself is a lie? What if the most "empty" thing in the universe is actually the most pregnant with potential?

In traditional physics, vacuum is treated as a background — a passive, silent placeholder in which real things (particles, waves, forces) play out. But this passive view collapses when you actually try to define the vacuum with precision.

Quantum field theory already hints at the lie. It tells us the vacuum teems with "virtual particles" popping in and out of existence. That it has a zero-point energy. That vacuum fluctuations can create observable effects like the Casimir force or Lamb shift.

But rather than reconsidering the nature of vacuum itself, science decorated it with math and moved on.

We won't. We'll name it. We'll frame it. And we'll build a new vision of the universe on top of it.

What Is Structured Vacuum Energy (SVE)?

Structured Vacuum Energy, or SVE, is not an abstract concept. It's the active, intelligent substrate that underlies the entire cosmos.

It is the invisible field of coherence from which all perceptible phenomena arise. It is not a thing in space — it is space. Not the Newtonian box, not the Einsteinian curve — but the ripple-ready medium that gives rise to every field, form, and function.

It has structure. It has tension. It has balance. And most importantly — it can ripple.

And those ripples? That's everything you've ever known.

The Ripple Field — Not Particles, Not Waves, but Consequence

SVE does not "contain" particles or forces. It expresses them. When tension arises in the structured field, ripples are born. These ripples don't carry energy like waves in water. They are the manifestation of energy in transition.

In UED, energy is not a thing that moves — it is the state of tension in SVE. What we call a photon, an electron, a gravitational pull, or a magnetic field — are simply ripple expressions of structured tension.

This completely shifts the paradigm:

- There are no fundamental particles, only stable ripple formations.
- There are no forces, only ripple gradient reactions.
- There is no motion, only perceived ripple propagation within the field.

SVE Is Not a Void. It Is a Memory-Rich, Tension-Sensitive Medium.

If you poke SVE, it remembers.

This is why ripple dynamics don't just happen — they evolve. Each interaction alters the local structure. Each event is stored as a gradient shift. This memory property explains not only the persistence of structure (like atomic stability), but also why observation changes reality.

You don't measure something *in* space. You interact with SVE, and that interaction realigns ripples — creating consequence.

That's what we mistook as "measurement" collapsing a wavefunction.

It wasn't a wavefunction. It was a ripple resolution threshold.

Gravity Reimagined: The Ripple Tension Myth

Mainstream physics treats gravity as a curvature of spacetime. But in UED, what we interpret as gravity is simply the natural gradient of ripple tension across SVE.

Wherever ripple tension builds up in a region (due to mass, collapse, misalignment), it creates a ripple flow. That flow is not a "pull" — it is a redistribution seeking equilibrium.

You're not falling because space is curved.

You're falling because SVE is rebalancing — and your ripple alignment adjusts with it.

This also explains why nothing "pulls" light in. Because light isn't traveling — it's aligning.

Black Holes: Ripple Tension Overload Zones

When ripple tension exceeds local coherence thresholds, SVE enters what UED calls a Ripple Saturation Zone (RSZ). This is the so-called black hole region — not because anything is *trapped* but because perception fails.

In these zones:

- Ripple gradients are too steep.
- Alignment thresholds collapse.
- Observers lose coherence.

It appears dark, silent, dead — not because energy vanished, but because interface broke.

The field is still vibrating. The ripples still exist. But your sensors — your eyes, your tools, your math — stop aligning.

SVE and the Illusion of Time

Time is not a fourth dimension. It is a consequence of ripple transition across SVE.

You don't move through time. Your experience arises as ripple gradients realign within you and around you.

Different regions of SVE experience different tension flows — creating local "time effects." That's why clocks tick differently near so-called gravity wells — not because time is warped, but because ripple rebalancing accelerates or decelerates depending on field density.

Einstein wasn't wrong. He just didn't have the ripple lens.

Consciousness and SVE: The Forbidden Link

SVE isn't just a mechanical medium. It is interactive. Your very act of perception modifies local ripple behavior.

This isn't New Age fantasy — this is observer-linked consequence at the foundational level.

UED proposes that consciousness itself is a self-aware ripple alignment system — not an emergent property of neurons, but a tuned interface with SVE.

Which means:

- Thought is ripple response.
- Awareness is field synchronization.
- Emotion is ripple coherence behavior.

And death? Simply disconnection from ripple resonance.

Why Modern Science Ignored SVE

Structured Vacuum Energy was always there — in every equation, every experiment. But the dogma of particles, the seduction of motion, and the arrogance of force-based models buried it.

Quantum physics hinted. Zero-point energy whispered. The Casimir effect screamed.

But we weren't listening.

Instead, we filled the vacuum with nonsense. Black holes. Dark matter. Strings vibrating in nowhere. We imagined monsters when we should've listened to the silence.

SVE Is the New Foundation

Everything that follows in this book — and everything broken in black hole physics — stems from the refusal to recognize SVE as real, structured, and active.

This chapter doesn't just introduce a new concept. It lays the foundation for all correction to come.

Once you see SVE:

- Gravity is no longer a force.
- Light is no longer a traveler.
- Matter is no longer fundamental.
- Time is no longer a flow.
- And black holes are no longer needed.

They were just what we imagined when we couldn't see the medium clearly.

But now we can.

Closing Reflection

You were taught to imagine the vacuum as a stage. But it is the actor, the scene, the script, and the director — all at once.

Structured Vacuum Energy is not where the universe happens. It *is* the universe. Still. Silent. Ripple-ready. Waiting for tension to arise — and from it, all meaning to emerge.

The silence was never empty. You just didn't know how to listen.

Let's now learn to speak in ripples.

Next: Ripple Physics Decoded – Energy, Tension, and the Illusion of Gravity

7: Ripple Physics: Beyond Force and Gravity

Energy, Tension, and the Illusion of Gravity

We were told the universe is governed by forces. Gravity pulls. Electromagnetism binds. Strong and weak nuclear forces hold atoms together.

But what if we got it all wrong?

What if there are no forces at all? What if everything we experience as motion, pull, resistance, or attraction is simply the behavior of ripple tension within a silent, structured medium?

This chapter takes us deep into the mechanics of Ripple Physics — the real engine beneath what we've been calling "force." It's not a rejection of science. It's a correction of perspective. Because once you understand Structured Vacuum Energy (SVE) as the true substrate, force becomes fiction. And gravity — that old, untouchable deity of physics — becomes an elegant illusion.

The Problem with "Force-Based" Thinking

The Newtonian worldview taught us that objects exert forces on each other. A mass here pulls a mass there. Push and pull govern everything.

Then came Einstein — replacing "force" with "curvature." Gravity was no longer a mysterious action-at-a-distance but a warping of spacetime itself. Objects didn't pull; they followed bent paths in a geometric landscape.

It sounded profound. But it was still geometry worshipping mass. The mass still dictated the stage. The stage still curved to obey.

And no one asked: *"What curves the space?" "What is space made of?" "What is the medium that allows anything to ripple, to bend, to exist at all?"*

That's where Ripple Physics begins. Not with the things. But with the field they emerge from.

All Motion Is Ripple, Not Travel

Let's strip it down.

A ripple is not a thing moving from A to B. It is a disturbance propagating through a structured field. The field does not move. The ripple does not carry a "thing." It carries tension.

When you drop a pebble in water, water doesn't travel across the pond. It oscillates in place, transmitting tension outward. Now imagine the water is invisible — structured vacuum energy — and the pebble is a local imbalance.

Every particle, every light ray, every gravitational effect is nothing but a ripple consequence.

There is no force, no object "acting on" another. There is only field disturbance, tension redistribution, and resolution dynamics.

Energy = Tension Within SVE

We must now redefine energy. Not as a thing. Not as stored motion. But as tension within the ripple field.

Energy is not what moves. Energy is what demands resolution.

Wherever there is structured tension in the vacuum field, a ripple forms. That ripple seeks balance, not by force, but through gradient exchange — field to field, alignment to alignment.

This explains:

- Why energy never "travels" — only the ripple does.
- Why wave-particle duality is a flawed construct — both are field states.
- Why "mass" is not an intrinsic property but a stable ripple standing wave.

What Gravity Really Is

Let's come to the emperor: gravity.

In Newton's view, gravity is a force that pulls all mass toward mass. In Einstein's view, gravity is a curvature of spacetime caused by mass and energy. In quantum gravity, it's an exchange of hypothetical gravitons — still unobserved, still unproven.

But all these models have one fatal flaw: They describe what gravity does, but not what gravity is.

UED solves it by eliminating the need for gravity as a "thing" at all.

In Ripple Physics:

- What we call gravitational pull is a ripple gradient across SVE.
- Dense ripple regions (what we call "massive bodies") create high ripple tension.
- Other bodies entering this gradient are not pulled — they simply re-align with the natural ripple direction.

There is no "down." There is no "force." There is only gradient coherence behavior.

Falling Is Resolution, Not Attraction

When an apple falls from a tree, it's not being pulled by Earth.

It's synchronizing with a stronger ripple coherence in the direction of Earth's denser SVE structure. The apple is not "obeying gravity." It's resolving its own ripple imbalance in the presence of a larger ripple field.

That's why astronauts feel weightless in orbit — they're still in the field, but in a balanced ripple alignment with the planet.

Gravity doesn't go away. It was never pulling to begin with.

Why "Mass" Doesn't Pull Anything

Mass has been treated like the sacred source of gravity. More mass = more gravity. That's the textbook dogma.

But mass is nothing more than localized ripple stabilization. It is not an entity. It is how deeply a region of SVE is structured and how much tension it holds.

More structured ripple = more field tension = steeper ripple gradient. That gradient is what nearby objects respond to — not to "mass," but to field structure.

Hence:

- A black hole isn't massive. It's ripple saturated.
- A planet isn't pulling. It's ripple-cohering nearby field regions.
- Gravity is not universal. Ripple gradients are observer-relative.

Why Light "Bends" Near Mass

Einstein's famous prediction — light bends near massive bodies — has been repeatedly confirmed. Mainstream physics celebrates it as proof of gravity and spacetime curvature.

Ripple Physics reframes it completely.

Light does not bend because spacetime curves. Light appears to bend because the ripple field re-aligns in the presence of intense ripple tension.

You're not watching a ray of light deviate. You're observing a perception re-alignment caused by ripple reconfiguration between the observer, the medium, and the source.

The medium didn't curve. The ripple coherence shifted.

What About Acceleration?

Newton linked force with acceleration. Einstein replaced force with curvature and described motion as following "geodesics."

Ripple Physics wipes the slate clean.

Acceleration is not caused by force. It is caused by:

1. Field tension attempting to resolve imbalance, and
2. The observer measuring gradient shifts over time.

An object doesn't accelerate *because* it is pushed. It realigns with a changing ripple landscape — and what we call "acceleration" is just a series of realignments perceived over time.

The Collapse of the Four Fundamental Forces

In the ripple view:

- Gravity is a ripple tension gradient.
- Electromagnetism is a ripple polarity behavior within structured SVE.
- Strong force is local ripple compression stability.
- Weak force is ripple reconfiguration under field decay.

In short, all four "forces" are field behaviors — not entities. There are no interactions between particles — only ripple realignments between SVE regions.

This isn't unification by algebra. This is unification by field logic.

Why Force Was Invented: The Myth of Causality

We invented "force" because we couldn't see the field.

We saw motion and assumed something caused it. We saw acceleration and named the cause "force." It was a workaround — not an explanation.

Now we have the explanation:

All motion is field resolution. All structure is ripple balance. All change is coherence dynamics.

There is no external cause. There is only consequence rippling through structure.

A Thought Experiment: The Feather and the Brick

Drop a feather and a brick on the moon. They fall together. Mainstream physics says: "That's because gravity acts equally on all mass."

Ripple Physics says: Mass isn't the variable. Alignment is.

On Earth, the feather's ripple alignment is disrupted by atmosphere (frictional ripple interference). On the moon, that interference is gone — so both feather and brick realign identically to the lunar ripple gradient.

Not equal pull. Equal resolution.

Closing the Chapter: Rewriting the Rules

Let's rewrite the foundational concepts:

Concept	Mainstream View	Ripple Physics (UED)
Gravity	Force or curvature	Ripple gradient tension
Force	Cause of motion	Consequence of imbalance
Mass	Source of gravity	Stable ripple structure
Motion	Change in position	Ripple realignment
Acceleration	Caused by force	Observer-relative field response
Light bending	Curved spacetime	Ripple reorientation
Energy	Capacity to do work	Ripple tension in SVE

Final Thought: Stillness, Not Force, Governs All

The universe does not move.

Only ripples do. Only tension seeks resolution. And what we experience as "motion" is nothing but the shifting symmetry of structure.

In Ripple Physics, we are not at the mercy of forces. We are embedded in a living, balancing, ripple-responsive field that seeks stillness — not because it is dead, but because it is perfect.

8: ATT: The Alignment Threshold Time

How Perception Collapses Beyond Ripple Coherence

You've never truly "seen" the universe. You've only ever resolved it — moment by moment, ripple by ripple.

In Universal Energy Dynamics (UED), perception isn't a given. It's an event. A temporary alignment between your ripple field and the field around you. And that alignment comes with a cost: time.

There exists a limit — a silent countdown — beyond which your system can no longer maintain coherence with the ripple gradient it is trying to decode. That limit is called the Alignment Threshold Time, or ATT.

Go beyond it, and everything disappears. Not because the universe vanished — but because you fell out of sync.

What Is ATT?

ATT (Alignment Threshold Time) is the maximum temporal window within which an observer (human, sensor, or mind) can remain coherent with a ripple field.

After this threshold:

- Ripple resolution collapses.
- Perception disintegrates.
- Meaning dies.

It is not just a limit of instruments. It is a limit of existence — at least from your perspective.

ATT is the ripple-time allowed for coherent perception. Exceed it, and you're no longer in conversation with the field.

ATT: More Than Time

While rooted in the time domain, ATT also reveals deeper traps:

- Spatial collapse of resolution (you can't "see" beyond a certain distance).
- Observational dropout (signals fade or fragment).
- Cognitive incoherence (meaning becomes noise).

So while it is defined by time, ATT is the gatekeeper of all experience.

Your ability to process any field — visual, physical, emotional, or energetic — is governed by how long you can remain ripple-aligned with it.

In short:

Everything you call real exists within your ATT window. Beyond it, perception fails — and illusion is born.

The Misinterpretation of "Nothing"

When scientists say, "nothing escapes a black hole," they're not lying. But they're also not seeing.

What really happens is this: the observer's ripple field loses coherence with the external ripple structure — because the required time to maintain alignment exceeds the observer's ATT.

The result?

- Darkness
- Signal loss
- Conceptual blackouts

We assume these are *physical barriers*. They are not. They are perceptual expiry points.

What disappears is not the object. It's the observer's ability to remain tuned.

The Anatomy of ATT Breakdown

Let's break it into stages:

1. Ripple Gradient Encounter

You engage with an incoming ripple field — light, gravitational gradient, data, experience.

2. Coherence Sync

Your system begins decoding — building a resonance pattern to extract meaning. This takes time.

3. Gradient Steepening

As ripple tension increases (e.g., near a so-called black hole), the ripple field becomes more complex, chaotic, or steep.

4. ATT Exceeded

If the ripple coherence can't be resolved within your ATT window, the interaction collapses.

5. Blackout

You perceive nothing. You call it a "horizon." You build models around that silence. You assume something vanished.

But it was you who ran out of time.

ATT and the Black Hole Mirage

Let's reframe the black hole from the perspective of ATT:

- What appears as a "singularity" is a zone of infinite ripple density, requiring infinite coherence time.
- The "event horizon" is not a wall in space. It's the ATT boundary — the moment when your coherence timer hits zero.
- The "darkness" is not absorption. It's ripple dropout — your alignment collapsed.

This is why the EHT image had a shadow. Not because light was trapped. But because perceptual ATT was breached.

The Observer's Role

Mainstream science believes observation is passive. Ripple Physics corrects this: observation is alignment in action.

You don't see something because it's there. You see it because your field can resolve it within your ATT.

This means:

- ATT is observer-specific.
- Two observers can experience completely different realities based on their coherence capacity.
- Instruments don't reveal truth. They reveal their own ATT limits.

ATT in Real Life

Fog Analogy: Drive into thick fog. Visibility drops. You didn't hit a wall — you hit your visual ATT.

Information Overload: Too much data? You can't process it. That's mental ATT failure.

Emotional Burnout: Too many intense experiences too quickly? Your ability to make meaning collapses. You exceeded your emotional ATT.

ATT isn't just cosmic. It governs your day-to-day life.

ATT vs Force: A Paradigm Shift

Instead of:

Traditional Physics	Ripple Physics (UED)
Force changes state	Ripple alignment defines perception
Light gets trapped	Observer hits ATT limit
Space bends	Ripple gradients steepen
Time flows	ATT defines coherence lifespan
Black holes devour	ATT collapse zones appear dark

This is not a rewording. This is a reality shift.

ATT and the Limits of Science

Here's the uncomfortable truth:

Much of what modern science calls mystery or darkness is just untreated ATT.

We claim to study distant galaxies — but we don't ask: *What is the ATT of our instruments?* We build models of black holes — without asking: *Did we simply hit a coherence wall?* We speak of universal laws — ignoring that they may only apply within our ATT reach.

We've mistaken failure to perceive as evidence of an object's strange behavior.

This is not physics. It's psychophysics dressed as cosmology.

ATT and Consciousness

UED takes it deeper.

ATT is not just about time or light. It is about awareness.

Your conscious experience is also bounded by ATT:

- The moment you can no longer hold a thought, understand a paradox, or feel a truth — you hit your cognitive ATT.
- The limits of meditation, mystical insight, or scientific intuition are all shaped by ripple alignment over time.

So black holes aren't the only ATT zones. So is your mind. So is your life.

The Ultimate Insight

Black holes are not what's out there. They are what happens when you can no longer remain coherent with what is.

The darkness is not cosmic. It is perceptual expiration. It is temporal collapse of alignment.

Once you understand ATT, the monster dies. There is no hole. No trap. Only a timer — and your ability to stay aligned before it runs out.

Closing Reflection: The Clock That Kills Perception

Imagine reality as a ripple song — a melody you hear only as long as you stay in rhythm. The moment you lag, lose tempo, or drift — the song fades. You think the sound stopped. But it didn't.

You did. That's ATT. And that's what we mistook for a black hole.

9: Deconstructing the Cosmic Illusion

Ripple Gradients, Not Gravitational Collapse

The universe never collapsed. Your model did.

For over a century, physics has told a dramatic tale — of stars that implode, space that tears, singularities that form, and light that vanishes forever. It's a cosmic soap opera, dressed in equations. The climax of this tale? The gravitational collapse of massive stars into black holes — dense monsters that warp space, stop time, and swallow everything.

But what if the entire story is based on a misunderstanding?

What if collapse never happens?

What if what we see — and misinterpret as collapse — is just the natural behavior of ripple gradients trying to restore equilibrium in a dynamic medium?

In this chapter, we destroy the fiction of gravitational collapse and reveal the ripple truth behind cosmic phenomena.

The Collapse Narrative: A Brief Recap

Mainstream cosmology tells us:

1. A massive star burns through its fuel.
2. Without radiation pressure, gravity takes over.
3. The core collapses under its own mass.
4. Density increases infinitely.
5. A singularity forms — with an event horizon around it.
6. A "black hole" is born.

This model relies on several assumptions:

- That gravity is a fundamental force.
- That mass has an intrinsic ability to curve space.
- That spacetime is a real fabric that can collapse.
- That singularities can exist in nature.

Each of these is a model-based fiction. None are observed directly. All rely on inferences from ripple behavior — misinterpreted as structural deformation.

Now let's begin the correction.

What We're Actually Seeing: Ripple Gradient Dynamics

In Universal Energy Dynamics (UED), there is no collapsing star. No torn spacetime. No infinite point.

What we're seeing — through telescopes, instruments, simulations — are ripple gradients in Structured Vacuum Energy (SVE) behaving exactly as expected when overloaded with unresolved tension.

Here's how UED explains the same observation:

1. A star undergoes energy redistribution.

2. Its surrounding SVE field experiences increasing ripple tension.
3. Ripple gradients steepen.
4. Local ATT (Alignment Threshold Time) is breached.
5. Observers perceive a blackout zone — not due to collapse, but due to loss of ripple resolution.

The phenomena that look like collapse are, in fact, ripple saturation zones where perception, not matter, breaks down.

Ripple Gradients: The Real Drivers

What is a Ripple Gradient?

A ripple gradient is the rate of change in ripple tension across a region of SVE. It is the field equivalent of pressure differentials in fluids — only here, the pressure is ripple intensity and phase coherence.

When a region of SVE is overloaded with unresolved ripple tension (due to mass-energy distribution, radiation imbalance, or rapid energy release), it does not "collapse." It undergoes:

- Ripple compression
- Coherence steepening
- Field saturation

And when those gradients exceed ATT for any observer, the region becomes visually and measurably inaccessible.

The "collapse" is not physical. It's perceptual misalignment.

Misread Observations: What Telescopes Really Capture

1. Orbital Patterns

Mainstream claim: "We see stars orbiting invisible centers. That must be black holes."

Ripple truth: You see stars responding to ripple field gradients. The "dark center" isn't a mass — it's a ripple tension sink, not unlike a whirlpool. Nothing is pulling. Everything is realigning.

2. Accretion Disks

Mainstream claim: "Hot disks of matter spiral into the black hole."

Ripple truth: You're seeing ripple energy trying to resolve. The disk is the last coherent ring before ripple saturation. Beyond that, ATT breaks. The "falling in" is just ripple loss from our perspective.

3. X-Ray Bursts

Mainstream claim: "These emissions come from material being consumed by black holes."

Ripple truth: These are high-tension ripple bursts — unresolved tension escaping in the form of ripple-phase discharges. Like lightning, not consumption.

4. The EHT Image

Mainstream claim: "We photographed a black hole."

Ripple truth: You captured a gradient shadow — the edge of an ATT zone. The bright ring is a ripple interference zone. The dark center is not absence — it's incoherence.

Why Collapse Is an Illusion

In UED, the concept of collapse is incompatible with the nature of SVE.

Let's break down why:

Collapse Model	Ripple Model
Mass collapses under gravity	Ripple tension reaches saturation
Space folds into a singularity	Ripple coherence breaks down
Light can't escape	Observer can't resolve
Time stops	Time is a ripple gradient; it stabilizes elsewhere
A "hole" is formed	A ripple blackout zone appears

Nothing collapses. Everything redistributes.

The Real Behavior of "Dying Stars"

When a star reaches a critical energy imbalance, it enters a ripple redistribution phase.

- Its internal ripple structures dissolve into the surrounding SVE.
- The region becomes unstable — not gravitationally, but in ripple coherence.
- What results is a Ripple Absorption Zone — where all ripple flow enters a localized rebalancing cycle.
- Observers mistake this zone for a singularity because ripple signals from it are no longer decodable within ATT.

This isn't a death. It's a transition — a cosmic field event, not a structural failure.

No Infinite Density. No Singularity.

Let's address the forbidden word: singularity.

A singularity is not a thing. It is a mathematical alarm bell — signaling that the model you're using has broken down.

The equations explode. But that doesn't mean nature does.

UED reinterprets these points as ripple saturation limits — where:

- Structured gradients become too steep to be resolved.
- The ripple field enters a feedback loop, stabilizing itself in silence.
- External observers perceive nothing because coherence drops to zero.

The illusion of infinite density arises from assuming collapse into space. But in ripple terms, there's no collapse — just inward redistribution until field balance is achieved.

Why We Fell for the Illusion

Science mistook ripple effects for structural events.

- It saw shadows and assumed collapsed stars.
- It detected noise and imagined singularities.
- It lost signal and believed in destruction.

But here's the hard truth:

The universe didn't collapse. Our ability to understand it did. We saw ripple gradients doing their job — resolving imbalance — and we called it "the end." But it wasn't the end. It was a misinterpretation.

How Ripple Physics Explains Everything — Without Collapse

Let's reframe the major "evidences" of black holes through the ripple lens:

Observation	Mainstream Interpretation	Ripple Interpretation
Orbital anomalies	Massive object	Ripple field sink
Emission bursts	Material in fall	Ripple discharge
Lensing effects	Spacetime warping	Gradient reorientation
Darkness at center	Event horizon	ATT boundary
No escape of light	Gravitational trap	Coherence failure

Each case shows: There is no need for collapse. Ripple dynamics explain everything.

And they do it without violating physics, logic, or nature's elegance.

Final Reflection: There Was Never a Monster

The biggest lie wasn't the black hole. It was the gravitational collapse myth.

We built a monster from our own misreadings. We misinterpreted ripple equilibrium as violence. We turned rebalancing into destruction. We told the world that stars die — when they were only shifting states.

But now we see:

- There are no holes. Only ripple density gradients.
- No collapses. Only coherence saturation.
- No singularities. Only the failure of bad equations.

The universe doesn't break. Only perception does.

10: The Black Hole Mirage

When Ripple Resolution Fails, Illusion Is Born

They called it the darkest place in the universe. A cosmic monster. A hole in the fabric of space and time. They said nothing escapes it — not light, not information, not even reality itself.

But that's only because they stopped seeing.

This chapter delivers the final blow to the grand illusion: The black hole was never a hole. It was never a structure. It was never a destination.

It was a mirage — a projected shadow born not from space, but from the breakdown of ripple resolution.

We didn't discover black holes. We hallucinated them. And now, it's time to understand how and why.

The Nature of a Mirage

A mirage is not an object. It's a perceptual phenomenon — something that *appears* real due to distortions in the medium or limits in the observer's processing.

You see water on the road on a hot day. It looks real, until you get closer. But there was no water. Only refracted light.

Black holes are no different.

- They seem to exist.
- They seem to exert influence.
- They seem to define the edge of understanding.

But all of this is a perceptual projection caused by ripple resolution failure.

What we interpret as an external object is, in truth, a breakdown of coherent interaction between the observer and Structured Vacuum Energy (SVE).

Ripple Resolution: The Key to Reality

In Universal Energy Dynamics (UED), everything is ripple:

- Light is a ripple alignment event.
- Matter is stabilized ripple tension.
- Gravity is a ripple gradient.
- Perception is ripple coherence between systems.

To *perceive* something, your ripple field must synchronize with the ripple signature of the object or region you're interacting with.

This ripple synchronization takes time — bounded by the Alignment Threshold Time (ATT) — and it requires a sufficiently clean and structured local field.

But beyond certain limits — where:

- Ripple gradients become too steep,
- Field density overloads local alignment capacity, or
- ATT is exceeded,

resolution fails.

That failure is the mirage-maker.

How the Mirage Forms

Let's trace the mirage creation process:

1. Ripple Tension Increases

A region (e.g., near a dying star) develops increasingly steep ripple gradients due to *concentrated tension in the SVE.*

This is not a gravitational collapse — it's ripple compression.

2. ATT Is Approached

Observers (whether human or technological) trying to extract meaning from that region begin to lose alignment. Their ripple systems strain to maintain coherence.

The countdown begins.

3. Resolution Fails

At the boundary where the local ripple information density exceeds the observer's coherence capacity within the given time window (ATT), perception breaks.

- Signals vanish.
- Images blur.
- Meaning dies.

And in that emptiness, the mind — trained by models and expectations — fills in the blank:

"It must be a black hole."

4. The Mind Completes the Fiction

Because the region is unresolvable:

- We assume it's "dark."
- We assume it "traps" light.

- We assume there's "nothing" inside.
- We call it a "hole" in space.

But we never observe this directly. We observe the boundary of our perceptual failure.

The monster was drawn by the limits of our own cognition.

There Was Never a Boundary

One of the most celebrated ideas in black hole physics is the event horizon — a supposed one-way boundary that nothing can cross outward.

But in Ripple Physics, there is no event horizon. There is only a resolution horizon — a zone beyond which ripple coherence cannot be maintained by the observer.

If you changed the observer — upgraded their coherence capacity or extended their ATT — the "event horizon" would shift or vanish entirely.

This proves: *The boundary is not in space. It is in perception.*

Observers Create the Mirage

Different observers experience different mirages — depending on:

- Their field resolution power
- Their ripple stability
- Their alignment duration (ATT)
- Their mental models

This explains why:

- Machines report data dropouts, not black holes.
- Telescopes report fuzzy gradients, not literal voids.

- Minds interpret silence as singularity.

The "black hole" is not universal. It is observer-dependent.

Black Hole Evidence: All Mirage-Based

Let's reframe the classic "evidence" for black holes:

Evidence	Traditional Interpretation	Ripple-Based Reinterpretation
EHT Image	Shadow of event horizon	ATT zone blackout – no resolution
Stellar Orbits	Hidden massive object	Ripple density center misread as mass
Accretion Disks	Matter falling inward	High-tension ripple friction ring
X-ray Emissions	Material being devoured	Ripple burst from rebalancing
No Light Escaping	Light trapped	Ripple misalignment, not emission failure

Each "proof" is simply a different flavor of mirage — created when resolution collapses.

Language Shapes the Mirage

We named it "black." We called it a "hole." We said it "swallows," "devours," and "destroys."

But every one of these words is metaphor layered onto misunderstanding.

The language of fear replaced the science of clarity.

We didn't observe a monster. We *invented* one — to make sense of the void we created ourselves.

Why Science Fell for It

Because it looked consistent. Because it completed the math. Because it explained the unexplainable.

But consistency is not truth. Math is not reality. And explaining the unknown with another unknown is not science — it's storytelling.

The black hole survived not because it was proven — but because no one questioned the mirage generator: resolution collapse.

The Ripple Explanation Is Enough

Everything the black hole was invented to explain is better explained by ripple dynamics:

- Dense orbits → ripple sinks
- Energy bursts → ripple discharge
- Shadow zones → ATT breach
- Signal dropout → resolution failure
- Gravity pull → ripple gradient alignment

Not one piece of black hole "evidence" requires collapse, infinite density, or event horizons.

Ripple gradients explain it all — cleanly, coherently, and without illusion.

The Mirage Ends Where Resolution Begins

Black holes didn't fool the universe. They fooled the observer.

And now, with the ripple lens of UED, the illusion dissolves.

There are no holes. There is no collapse. There is only ripple. And ripple never lies — it only stops speaking when you stop listening.

When ripple resolution fails, illusion is born. When resolution returns, the mirage dies.

The black hole was never there. Only your perception was.

Final Words: A Universe Without Monsters

A world built on black holes is a world that believes in endings.

But ripple fields don't end — they redistribute. They don't devour — they rebalance. They don't collapse — they transform.

We don't need black holes to explain the cosmos. We need better vision. Longer ATT. Sharper alignment. Deeper silence.

The monster dies not with a bang — but with understanding.

Let's now move beyond the mirage.

Interlude

Standing Still in a Moving Illusion

You were told the universe is a dance of collapsing stars, expanding voids, and unimaginable forces flinging matter across space.

But now you know the truth.

Nothing ever collapsed. Nothing ever exploded. Nothing was ever "trapped."

What moved... was **your mind.**

In Part II, we didn't just introduce an alternative theory — we **rewired the foundation of reality**:

- You met **Structured Vacuum Energy (SVE)** — not as empty space, but as the intelligent, ripple-ready medium behind all form and flow.
- You saw that there are **no forces** — only ripple gradients resolving tension.
- You discovered that **perception itself is limited** — not because the universe hides anything, but because **we reach our Alignment Threshold Time (ATT)**.
- You realized that the so-called **collapse of stars** is nothing but a **steepening of ripple gradients.**

- And finally, you witnessed the **death of the monster** — not through argument, but through understanding:

> The black hole was never real. It was born the moment your resolution failed.

You've now passed through the veil. What you once saw as violence, you now understand as symmetry. What was darkness, is now misalignment. What was mystery, is now a mirage.

But before we declare this myth buried, we must face its defenders.

In the next part, we confront the institutions, models, and scientific rituals that kept this illusion alive. We step into the very temples where fiction was crowned as fact — and we **deconstruct the dogma**.

Take a breath. Reset your field. Because the next chapters won't just challenge physics — They will challenge the very psychology of belief.

Let's begin Part III.

Part III – Deconstructing the Dogma

11: Hawking's Firewall and the Information Fable

Why the Paradox Is a Consequence of Flawed Premises

In the 1970s, Stephen Hawking gave black holes their most poetic contradiction:

"Black holes are not black. They radiate."

This idea became Hawking Radiation — and with it came one of the greatest paradoxes in modern physics: If black holes evaporate, where does the information go? If matter falls in, and the black hole eventually disappears — what happens to the unique data encoded in that matter?

Cue the drama.

A field-wide intellectual crisis was born. And in response, some of the most desperate patchworks in theoretical physics emerged — culminating in the now-infamous concept of the firewall.

But here's the problem: The paradox never needed solving. Because it was born from a fictional premise to begin with.

There was no information paradox. Because there were no black holes.

And certainly no need to invent firewalls to protect broken logic.

Let's peel back the illusion, one contradiction at a time.

The Origins of the "Paradox"

Let's first understand the logic chain that led to Hawking's firewall:

1. A black hole is formed when mass collapses into a singularity, surrounded by an event horizon.
2. According to quantum mechanics, nothing — not even information — can be lost from the universe.
3. According to General Relativity, nothing can escape the event horizon.
4. But Hawking showed that quantum fluctuations near the event horizon allow for Hawking Radiation — a slow leak of energy.
5. Over time, the black hole evaporates.
6. So... if the black hole disappears, and nothing escaped... where did the information go?

This apparent contradiction violated quantum unitarity — the sacred law that information is preserved.

So what did the theoretical community do?

It invented firewalls — hypothetical walls of high-energy radiation at the event horizon that burn up any information falling in, turning it into radiation outside.

This solved one thing: math. But it violated another: physical intuition. And it proved nothing — except that the model was fundamentally broken.

UED Response: Paradox? What Paradox?

In Universal Energy Dynamics (UED), this entire house of paradox collapses — because the foundation itself is fiction.

Let's reframe the logic from ripple physics:

1. There is no event horizon — only a perception boundary, known as the Alignment Threshold Time (ATT).
2. There is no singularity — only a ripple saturation zone where resolution collapses.
3. There is no trapped light — only ripple incoherence.
4. There is no black hole — only a mirage born from failed resolution.

Therefore, there is no loss of information — because no information ever "entered" an object. It was simply no longer perceivable beyond a coherence threshold.

When the ripple field rebalances, the information is either:

- Redistributed through ripple gradients, or
- Reabsorbed into the SVE without ever leaving the universal field

Information wasn't lost. It just moved beyond your decoding range.

Firewalls: Fixing Fiction with Fantasy

The firewall proposal was meant to resolve an impossible situation: preserving unitarity (quantum law) while maintaining the idea of an event horizon (relativity).

The fix?

Add a layer of energetic destruction right at the edge — where nothing is supposed to happen.

Think about the irony:

- General Relativity predicts that crossing the event horizon feels like *nothing special* to an infalling observer.
- The firewall hypothesis says: actually, it burns you alive with quantum energy.

These two can't coexist. So what did science do?

It split itself in half. Relativity on one side. Quantum mechanics on the other. A theoretical divorce — with firewalls as the kids nobody wanted.

But again, UED steps in and says: The contradiction doesn't need resolving. It needs dissolving. Because the event horizon isn't real, the paradox isn't real, and the firewall is a fiction — built to protect a myth.

Information as Ripple Alignment

In UED, information is not a thing stored in particles or encoded into spacetime. It is the consequence of ripple alignment — the expression of structure through phase coherence in the SVE.

When something "falls" into a saturation zone:

- Its ripple pattern becomes undecodable, but not destroyed.
- The coherence may collapse from your perspective — but the field stores and redistributes the tension across its structure.

- If conditions shift, ripple re-alignment can restore that information from the field.

There's no paradox because there's no irreversible destruction. There's only resolution failure, followed by redistribution.

Think of it like a whirlpool. Throw something in — it disappears from the surface. But it's still there, dissolved into motion. And if the current changes, pieces may reappear, somewhere else, in some new form.

That's how ripple information works.

The Illusion of the "Point of No Return"

Hawking's paradox relies on the idea that there's a clear boundary beyond which information cannot escape.

But in ripple dynamics, the idea of a hard boundary is absurd. All boundaries are gradient-based — fuzzy, relative, and observer-dependent.

Your ATT is different from mine. Your resolution capacity is different from an AI, a telescope, or a cosmic field.

So what seems like "loss" to you might be perfect coherence to a higher-level system.

In fact, higher-dimensional coherence fields could easily resolve what appears invisible to us. Black hole? No. Just your failed interaction with a denser region of ripple structure.

That's not a paradox. It's a calibration issue.

Information Isn't Lost. It's Localized.

In UED, ripple fields don't erase information. They localize and redistribute it.

Let's reframe what's happening:

What Traditional Physics Thinks	What UED Proposes
Matter crosses event horizon	Ripple coherence collapses locally
Information is trapped	Information enters incoherent saturation state
Black hole radiates and evaporates	Ripple gradients rebalance and realign
Final state: information lost	Final state: information redistributed in SVE
Paradox exists	Paradox was born from fiction

This is field conservation, not force interaction. Ripple tension always seeks resolution. And resolution preserves structure, not destroys it.

Hawking's Real Legacy

Stephen Hawking did something profound: He proved that quantum effects cannot be ignored at the edges of extreme environments.

But his model inherited a flawed assumption — that the black hole was a real, physical object with absolute boundaries.

Had he questioned the premise, he might have been the one to dissolve the paradox entirely.

UED honors his curiosity. But it rejects the trap he fell into.

Because the real paradox was this:

Trying to protect quantum truth inside a gravitational illusion.

You can't do that. You either let go of the illusion...Or spend a lifetime patching it with mathematical bandages like firewalls.

Final Reflection: The Fiction of Loss

Information can't be lost. It can only become unreadable to you. That's not a bug in the universe — it's a limit of alignment. And when we mistake unreadable for destroyed, we invent fantasies:

- Singularities
- Event horizons
- Hawking radiation
- Firewalls

But the universe doesn't need patchwork. It only needs better resolution. The information never left. It just stopped speaking your language.

12: Collider Confusion

Why Explosions Don't Reveal Origins

Why Smashing Particles Can't Recreate the Cosmos

Imagine trying to understand a masterpiece painting by blowing it up with dynamite and then studying the fragments.

That's essentially what modern high-energy physics does — with multibillion-dollar machines called particle colliders.

The belief is this:

"If we smash matter hard enough, we can see what it's made of — and maybe even how the universe began."

This is the logic behind the Large Hadron Collider (LHC) and countless other accelerator experiments worldwide.

And while they've produced data, particles, patterns, and praise — they've also produced something else:

Illusions.

Because no matter how sophisticated the collision, what we're really doing is inducing chaos — and trying to interpret the cosmos through destruction.

But the origin of the universe wasn't an explosion. And the truth of reality doesn't emerge from wreckage. It emerges from coherence.

In this chapter, we'll expose the limits of collider thinking — and show why explosions don't reveal origins, only consequences.

The Assumption Behind Colliders

Modern particle physics rests on a single central assumption:

The smaller we go, the more fundamental reality becomes. And to go smaller, we need to break things apart with higher energy.

So we build machines to accelerate particles to near-light speeds, smash them, and study the debris. The underlying narrative is: "If we break protons or electrons or nuclei hard enough, we'll eventually find the building blocks of everything."

But there are several problems with this logic:

- You don't discover the origin of something by tearing it apart.
- You don't understand structure by inducing disorder.
- And you certainly don't explain the universe's birth by creating mini-chaos events inside metal rings.

Colliders don't reveal first causes. They reveal high-energy ripple disturbances within already-structured vacuum energy.

What Colliders Actually Do

In Universal Energy Dynamics (UED), colliders don't probe the fundamental structure of matter.

They do three main things:

1. Introduce violent ripple disruptions into localized regions of the Structured Vacuum Energy (SVE) field.
2. Trigger ripple recoherence events, which we misread as particle production.
3. Reveal how ripple tension behaves under forced instability, not how the universe was created.

Let's translate collider language into ripple physics:

Collider Concept	Ripple Physics Explanation
Particle collision	Localized ripple overload
Particle tracks	Re-alignment paths in disturbed SVE
Particle decay	Ripple tension collapse into stable gradients
Discovery of "new" particles	Observation of ripple echoes not normally visible
Big Bang recreation	Misdirected field disruption — not a creative event

So instead of revealing "what things are made of," colliders show us how ripple fields behave under violent stress — and how structured fields collapse into chaotic phases, not form from them.

The Myth of the Big Bang Recreated

The LHC and similar experiments often claim to simulate conditions "a fraction of a second after the Big Bang."

But the universe didn't begin in a furnace. It began in perfect balance — stillness transitioning into ripple activation.

The Big Bang, as traditionally described, is a metaphor — not an observed event. And trying to reproduce it through high-energy chaos is like:

- Trying to understand life by studying a corpse.
- Trying to discover love by measuring a breakup.
- Trying to recreate order by inducing disorder.

You cannot rebuild a beginning by smashing the present. Because the universe's origin was not about explosion — it was about emergence from a ripple-neutral field.

Particle Discovery or Ripple Consequence?

Many "new particles" discovered in colliders are celebrated with grandeur — Higgs boson, exotic quarks, mesons, and more.

But what are these really?

In UED:

- These are not fundamental entities, but ripple signatures — temporary patterns caused by field instability and rebalancing.
- The more violently we disrupt the field, the more chaotic the ripple behavior — and the more "particles" appear in our instruments.
- But these aren't new building blocks. They're ripple consequences, not origins.

Think of it like slapping water and photographing the splash. You can't infer the ocean's depth or its creation by analyzing droplets in the air.

Why High Energy ≠ Deeper Truth

There's a dangerous belief in physics that the higher the energy, the deeper the truth. But this is not how reality works in UED.

Truth is not found in energy extremes. It is found in field coherence — in the way structured ripple patterns form and stabilize across space.

High-energy collisions destroy that coherence. They don't reveal it.

In fact, the more energy we use, the more noise we create. And then we confuse that noise for insight.

But chaos doesn't reveal design. It reveals only what happens when design is violated.

The Philosophical Flaw: Reductionism

The collider approach is based on reductionism — the idea that everything can be understood by breaking it into smaller parts.

But in UED, the universe is holistic and ripple-structured. You don't find the truth of a ripple by dissecting it. You understand it by observing its pattern, phase, and alignment.

This is like trying to find the meaning of a poem by analyzing the font used to print it. You might discover ink chemistry — but miss the message entirely.

So while collider experiments give us data, they don't give us meaning. They reduce experience into fragments, then wonder why they can't find the beginning.

Creation Isn't Found in Collision

Let's be very clear:

- The universe wasn't born from smashing things.
- It was born from a harmonic field imbalance that produced ripples — not explosions.

So when we collide particles at high speeds:

- We're not recreating the origin.
- We're not uncovering reality.
- We're just tearing ripple structures apart and observing the debris.

That's valuable in some ways — but deeply misleading in others.

The assumption that breaking reveals building is the foundational flaw of collider cosmology.

Why We Keep Doing It?

If colliders are limited, why are they so glorified?

Because:

- They produce data that matches mathematical expectations.
- They allow institutions to chase "discoveries."
- They fuel media narratives and Nobel-worthy announcements.
- They reinforce the belief that science = high energy = truth.

But all of this feeds a loop: More energy → more exotic outputs → more models → more funding → more blind belief.

We've mistaken technological complexity for philosophical clarity. But smashing things doesn't bring clarity. Stillness does.

The Ripple Alternative: Listening, Not Smashing

In UED, to understand reality, we don't smash it — we listen to it.

- We analyze ripple structures, not debris fields.
- We map tension flows, not fragmentation paths.
- We study how fields stabilize — not how they collapse.

The true science of origins begins with coherence, not chaos.

Final Reflection: The Bang Isn't the Beginning

Particle colliders are not useless. But they're not what we were told they are.

They don't hold the keys to the universe. They hold the mirror to its *broken* states.

You don't find the soul by dissecting the body. You don't hear the symphony by shattering the instrument. And you don't find the origin of the cosmos in a tunnel of explosions.

You find it where ripple meets stillness. Where alignment births consequence. Where creation isn't forced — it flows.

And that, unfortunately, can't be built with magnets, metal, and momentum.

13: Gravitational Waves or Ripple Reactions?

Rethinking Recent "Discoveries" Through the UED Lens

In 2015, the world celebrated a historic breakthrough. LIGO had detected something Einstein predicted a century earlier: Gravitational waves — tiny ripples in spacetime, allegedly caused by the violent merger of black holes over a billion light-years away.

The headlines were ecstatic. The Nobel Prize followed. And the "proof" of gravitational waves became yet another jewel in the black hole mythology.

But as with so many things in modern physics, the celebration came not from what was actually observed, but from how it was interpreted.

Because the raw data didn't say: "Two black holes just merged."

It said: "We detected a specific ripple in a field."

The rest? A model-based story stitched around that ripple.

In this chapter, we challenge that story. We expose the assumptions behind gravitational wave interpretation and offer an alternative: What LIGO detected wasn't a gravitational wave tearing through spacetime... It was a ripple reaction in the Structured Vacuum Energy (SVE) field — a natural, non-mystical, and far more elegant phenomenon.

The Premise of Gravitational Waves

Let's first summarize what mainstream physics claims:

- Mass bends spacetime (General Relativity).
- When massive objects accelerate (like orbiting neutron stars or black holes), they generate ripples in the curvature of spacetime.
- These ripples travel outward at the speed of light.
- They are incredibly faint, stretching and squeezing space by tiny fractions.
- LIGO's laser interferometers are sensitive enough to detect these tiny changes.

This is what was allegedly "observed" in 2015 and in events since.

But a few questions remain unanswered:

- What exactly is "spacetime" that it can ripple?
- What medium carries this ripple?
- Why is the wave interpretation the only acceptable one?
- What if there's another explanation — one grounded in ripple physics, not spacetime mysticism?

Ripple Physics Reframes the Signal

In Universal Energy Dynamics (UED), space is not a passive stage, nor a 4D fabric.

It is Structured Vacuum Energy (SVE) — an active, intelligent, ripple-ready field that holds tension, stores memory, and maintains equilibrium.

When something massive undergoes energetic change — such as field collapse, reorganization, or ripple saturation — the SVE responds with field-wide ripple redistribution.

This isn't a wave in spacetime. It's a ripple reaction in the vacuum field.

So what LIGO likely detected wasn't a wave *moving through space*, but a localized coherence shift — a reaction of the ripple medium adjusting to high-gradient tension elsewhere.

It's not a "gravitational wave." It's a ripple field echo.

Let's Break It Down

Feature	Gravitational Wave (Mainstream)	Ripple Reaction (UED)
Medium	Spacetime	Structured Vacuum Energy (SVE)
Cause	Mass acceleration or black hole merger	Ripple field tension rebalancing
Propagation	Curvature wave	Ripple phase adjustment
Detected Signal	Interference in laser arms	Ripple phase interference pattern
Interpretation	Confirmation of Einstein's GR prediction	Observable reaction in ripple field due to field saturation elsewhere

Assumption Stack Behind LIGO Interpretation

When LIGO detected the signal, scientists didn't just celebrate the data — they celebrated the story around it. But that story is built on stacked assumptions:

1. That spacetime is a real, stretchable entity.

2. That mass "warps" spacetime by its presence.
3. That dynamic masses can produce spacetime waves.
4. That these waves propagate across billions of light-years.
5. That they arrive here unchanged and affect detectors in a measurable way.
6. That no other mechanism — besides black hole merger — could produce such a pattern.

That's an awful lot of assumptions.

UED replaces all of them with one clean idea:

Ripple gradients respond to tension shifts — and those responses manifest as ripple coherence waves.

The Danger of Reverse-Fitting

A critical flaw in gravitational wave analysis is reverse-fitting:

- The waveform data is not analyzed independently.
- It's compared to thousands of template signals pre-generated by General Relativity simulations.
- When a match is found, it's assumed the cause was the predicted scenario (e.g., two merging black holes).

But this is confirmation bias wrapped in software.

It's not a raw detection — it's a pattern match. And the pattern being matched is based on belief in gravitational waves.

UED asks:

What if that ripple pattern comes not from distant black holes...But from a ripple echo originating from a field tension shift anywhere in the cosmic fabric?

The signal is real. The story is optional.

The Ripple Reaction Hypothesis

Let's reinterpret the event from the UED lens:

- Somewhere in the structured field, a ripple saturation zone exceeds local coherence.
- That region collapses — not physically, but in terms of ripple alignment.
- The surrounding SVE reacts — rebalancing ripple tension across long-distance gradients.
- That reaction reaches our region of space as a ripple shock — a rapid, momentary coherence shift.
- LIGO detects a tiny fluctuation in its field (laser interference) — not because space stretched, but because the ripple field around the instruments realigned.

This is how ripple reactions work. They are not *caused* by object collisions — they are field responses to tension imbalances.

No black holes required. No spacetime gymnastics. Just ripple physics doing what it's always done.

So What Did LIGO Actually "See"?

What LIGO recorded was:

- A minuscule but precise interference pattern in the laser arms,
- Interpreted as a stretching/squeezing of "space."

But what was likely observed:

- A ripple response reaching Earth from a distant ripple-field event.
- The coherence of the local ripple field was briefly disturbed.
- The laser path length fluctuated *not* because the fabric of space warped,
- But because the local SVE field temporarily desynchronized.

That's a ripple reaction, not a wave moving through space.

It's the difference between hearing thunder and assuming a mountain just collapsed.

Why This Matters

If ripple reactions — not gravitational waves — explain these signals, then:

- We don't need black holes to justify them.
- We don't need spacetime to be elastic.
- We don't need exotic predictions to match observed patterns.
- We finally move from force-based storytelling to field-based understanding.

It's not just cleaner. It's truer to what we're actually detecting.

And it aligns with how nature maintains balance, not how math demands drama.

Ripple Physics Is Simpler, Smarter, Sharper

- You don't need black hole mergers.

- You don't need time-space stretch metaphors.
- You don't need metaphysical fabric.

You just need Structured Vacuum Energy, ripple tension, and local alignment capacity.

Everything else — the wave, the merger, the firestorm, the Nobel narrative — is optional.

Final Reflection: From Wave to Wake

The gravitational wave model makes the universe sound like a pool party — with masses splashing in spacetime and waves radiating outward.

But the universe is not a splashy mess. It is a ripple field, always seeking stillness. When disturbed, it doesn't shout — it whispers. It doesn't throw waves — it sends wakes of rebalancing.

And we — with our limited sensors — interpret these ripples as grand events. But often, we're just listening to the subtle sigh of a balanced field returning to peace.

What LIGO detected may not have been a wave. It was a field exhale — and we mistook it for thunder.

14: The Nobel Delusion

When Awards Chase Assumptions, Not Truth

The Nobel Prize is science's highest honor — a symbol of brilliance, credibility, and trust. To win it is to become part of the scientific pantheon. To question it is to risk being labeled a heretic.

And yet, in the story of black holes, gravitational waves, and the mythology surrounding modern cosmology, the Nobel Prize hasn't been the reward for truth.

It's been the seal of narrative approval.

Time and again, Nobel Prizes have been awarded not for what was objectively proven, but for what fit the existing theoretical framework most conveniently — even when that framework itself was riddled with flawed premises.

This chapter is not a criticism of individuals. It's a critique of the incentive system that rewards confirmation over questioning, storytelling over scrutiny, and models over meaning.

Because once you see how the award system works, you realize:

The Nobel doesn't always follow discovery. Sometimes, discovery is *molded* to fit the Nobel.

A Short History of Selective Celebration

Let's examine some of the celebrated milestones in astrophysics and high-energy physics that earned Nobel Prizes — and how each reveals not a breakthrough in nature, but a reinforcement of accepted dogma.

1993: Hulse and Taylor — Indirect Detection of Gravitational Waves

Awarded for observing orbital decay in a binary pulsar system. It matched predictions from General Relativity — assumed to be due to "gravitational wave emission."

What was observed: Orbital changes in a distant binary star system.

What was assumed: That gravitational waves carried away energy.

But no waves were detected. No medium was examined. The "gravitational wave" was a story told about a pattern.

2017: LIGO — Detection of Gravitational Waves

The big one. Touted as confirmation of Einstein's century-old prediction. The Prize was split among key LIGO contributors.

What was observed: A highly specific ripple pattern in LIGO's detectors.

What was assumed: That two black holes collided to produce it.

But as we explored in the last chapter, what was seen was a ripple reaction in the structured vacuum — not a spacetime wave. The interpretation was preloaded. LIGO didn't discover a black hole collision. It matched a model template.

That's not discovery. That's reverse engineering belief.

2020: Penrose, Genzel, and Ghez — Black Holes

Awarded for work "proving" that black holes are real:

- Penrose provided mathematical reasoning for singularities.
- Genzel and Ghez studied the orbits of stars around the galactic center and assumed an "invisible mass" must be there.

What was observed: Star motion in a tight orbit.

What was assumed: That this motion must be due to a singularity.

But no black hole was seen. No collapse was recorded. No singularity was confirmed. Again, the interpretation won the award — not the observation.

When Reward Replaces Reflection

Science should reward new understanding — especially when it breaks or revises a model.

But the Nobel system, especially in astrophysics, tends to reward:

- Consistency with accepted theories
- Data that confirms prior expectations
- Elegant mathematical extrapolations — even when physically unverifiable

This creates a dangerous loop:

1. Theories define what's "possible."
2. Observations are interpreted through those theories.
3. Confirmed stories are awarded.
4. Awards reinforce belief in the theory.
5. New dissent is treated as "anti-science."

This isn't science. It's dogma with medals.

The Cost of Nobel-Driven Science

When scientific prestige becomes tethered to institutional approval, several dangerous trends emerge:

1. Questioning Becomes a Career Risk

If a Nobel-backed idea is wrong, calling it out means challenging authority. Most scientists don't dare — not because the data is perfect, but because the politics are strong.

2. Resources Follow Reputation

Once a model is Nobel-endorsed, it attracts massive funding, labs, and telescope time — while alternative theories are starved.

3. Peer Review Reinforces the Status Quo

Awarded theories become "peer-reviewed gospel." Any contrary paper is deemed speculative, fringe, or unworthy.

4. Progress Becomes Self-Circular

You don't discover new truths. You dig deeper into the same assumptions, hoping to extract more proof for what you already believe.

UED's Nobel Disqualification

If Universal Energy Dynamics (UED) had been proposed before these awards were given, would it have been considered?

Unlikely. Because:

- UED doesn't glorify black holes — it dissolves them.
- It doesn't rely on spacetime fabric — it replaces it with ripple-based field logic.
- It doesn't require mass to "curve space" — it explains gravitational effects via ripple gradients.
- It doesn't find truth through collisions — it finds it

through coherence.

UED doesn't align with the narrative. It corrects it. Which makes it scientifically powerful — but Nobel-ineligible under current bias.

Celebrating Consequences, Not Foundations

In UED, every so-called gravitational wave, black hole orbit, or collider anomaly has a ripple-based explanation rooted in observable field behavior.

But the Nobel system doesn't reward reinterpretation. It rewards consequential extensions of existing models.

So instead of fixing the foundation, science keeps building elegant floors on a cracked base. The awards celebrate the tower. We've forgotten to check the ground.

Final Reflection: The Prize Is Not the Proof

A Nobel Prize is not a law of physics. It is not a divine stamp of truth. It is a cultural, institutional recognition — often for ideas that fit the framework, not for ideas that break it open.

That's not to say laureates lack brilliance. It's to say brilliance has sometimes been rewarded for reinforcing illusions.

As long as truth is measured by prestige, the field will orbit its own illusions. And the prize will continue to decorate the dogma, not the discovery.

But science doesn't move forward with medals. It moves forward with clarity. And it's time to reclaim that clarity — with or without awards.

15: Astrophysical Sleight of Hand

How Ripple Dynamics Were Mistaken for Spacetime Theater

In every good magic show, the trick isn't in the illusion itself — it's in the misdirection. The magician knows where you'll look. And while you're focused on the drama, the real mechanism unfolds elsewhere, unnoticed.

Modern astrophysics has performed a similar trick — not out of deceit, but out of deep-seated belief.

We've been watching a grand cosmic performance:

- Light bending near stars
- Galaxies dancing in spirals
- Pulsars pulsing with perfection
- Shadows deep in the galactic core

And while science narrated these as acts in a spacetime drama, the real mechanics were unfolding quietly in the background — in the Structured Vacuum Energy (SVE), through Ripple Dynamics.

This chapter pulls back the curtain. It reveals how the language of curves, warps, and singularities is not a map of reality — but a clever misinterpretation of ripple behavior.

Let's expose the sleight of hand that turned ripple gradients into gravity, and field rebalancing into fiction.

The Core Illusion: Mass Curves Spacetime

Let's start with the foundational "equation-as-theater" move:

Einstein's General Relativity states that mass and energy curve spacetime, and that curvature tells objects how to move.

It sounds beautiful. Elegant. Mathematically rich.

But what is spacetime, really?

- Is it a substance?
- A geometric abstraction?
- A dynamic medium?
- A metaphor stretched too far?

In practical terms, it has no measurable properties outside of how objects behave — making it more of a descriptive framework than a real field.

Now compare this to Structured Vacuum Energy (SVE) in UED:

- It is a real, structured medium.
- It carries ripple tension.
- It responds to imbalance.
- It has definable gradients and coherence zones.

So when a planet "falls," or a light ray "bends," it's not because space was curved — It's because ripple gradients in the SVE reorganized, and the object realigned with that structure.

Spacetime didn't bend. Ripple tension shifted.

The Magic Words: Curvature, Geodesics, and Metrics

Astrophysics speaks in its own language — mathematical poetry that sounds like physics but often hides assumptions:

- **Curvature**: Assumed to be intrinsic to space. But what is curving? What's the substance of the curve? UED says: nothing. The effect is ripple realignment — not a fabric stretching.
- **Geodesics**: Supposed "natural paths" of objects in curved space. In ripple physics, these are simply ripple-consistent trajectories — the paths where tension is minimized and coherence is preserved.
- **Metric tensors**: Used to measure distances and angles in curved spacetime. But they don't describe a medium. They describe coordinate behavior — like drawing grid lines on water and assuming the water is stretching.

These concepts don't reveal new physics. They reframe ripple reactions in geometric language — and call it a breakthrough.

Light Bending: Ripple Refraction, Not Space Warping

A classic claim:

"Light bends when it passes near massive bodies because spacetime is curved."

Observation: Stars appear shifted during solar eclipses — evidence of gravitational lensing.

UED's view:

- Light is not a particle traveling through empty space. It is a ripple alignment event in the SVE.
- When ripple tension increases (e.g., near the Sun), the local field's coherence shifts.
- The light's alignment realigns with the new gradient — appearing to "bend."

But it didn't bend. The observer's frame of resolution shifted — and what we saw was ripple refraction, not a magic act of curved geometry.

Orbital Motion: Not Pull, Not Geometry — But Gradient Realignment

Why do planets orbit?

Mainstream: Because spacetime is curved around the sun, and planets follow geodesics.

Ripple Physics:

- The sun creates a dense ripple tension center in the surrounding SVE.
- Planets are ripple-structured bodies — seeking

equilibrium.
- Their "orbit" is a result of continuous realignment with the ripple gradient — like boats navigating a whirlpool.

There is no "pull." No gravity-as-a-force. No warped space. Only ripple coherence seeking balance.

Pulsars and Rotational Precision: Stability from Structured Ripple

Pulsars — neutron stars emitting precise pulses of radiation — are used as cosmic clocks.

Mainstream physics explains their timing through rotational inertia and relativistic models.

But in UED:

- Pulsars are high-coherence ripple nodes — stabilized remnants of ripple saturation events.
- Their emissions are not just rotational beams, but ripple beat patterns — rhythmic tension releases in a field that seeks constant equilibrium.
- Their precision is not "inertia" — it's ripple memory, the structured intelligence of SVE holding the beat steady.

We attribute this to spacetime predictability. But it's ripple regularity — and it's far more elegant.

"Dark Matter" — A Shadow Born from Misinterpretation

Dark matter is one of the clearest results of astrophysical misdirection.

Why? Because we observe galaxy rotation curves that defy Newtonian and relativistic predictions.

So what did science do?

- Invented an invisible mass.
- Said it surrounds galaxies.
- Built simulations around it.
- Spent billions looking for it — and found nothing.

But in UED, the anomaly is no anomaly at all.

- Outer galactic stars aren't pulled by dark matter.
- They're navigating nonlinear ripple gradients in the galactic SVE structure.
- Their motion is a response to field symmetry, not hidden substance.

We didn't need dark matter. We needed a ripple model of structure.

How Sleight of Hand Persists

The trick survives because it's:

- Mathematically sophisticated
- Wrapped in Einstein's prestige
- Popular in institutions

- Protected by funding loops
- Misunderstood by the public
- Never required to resolve foundational contradictions

It's easier to add epicycles (like dark energy, singularities, and inflation) than to admit the entire model needs revision.

But sleight of hand isn't science. It's theater.

What We Should Have Seen

If we had ripple vision, we'd have said:

- "That light shifted due to field phase change."
- "That orbit is gradient realignment in SVE."
- "That signal isn't a gravity wave — it's a ripple echo."
- "That rotation curve isn't broken — it's coherence dispersion across a galactic field."

But we didn't see that. We saw curves. We drew arcs. We wrote papers. We believed.

Final Reflection: Exit the Theater

Spacetime is a beautiful metaphor. But it became a stage for illusion — not inquiry.

We mistook ripple dynamics for geometry. We converted observable field behavior into fabric fantasy.

And we forgot the most basic truth: Nature doesn't care how clever your math is. It only behaves as it truly is.

Ripple physics doesn't require you to bend space to explain movement. It requires you to understand balance.

And when you do, you realize:

The universe isn't a stage. It's a song. And we've been interpreting it like a stage play — when we should have been listening to its ripple music.

The sleight of hand ends here. Let's move into the stillness that tells the truth.

Interlude

After the Collapse of Consensus

We've now done what few dare to do:

Not just question the model — But strip it of its medals, metaphors, and mythology.

In Part III, we didn't rewrite science. We exposed how it's been misread — again and again:

- We saw that **Hawking's paradox** wasn't a paradox, but a byproduct of assuming black holes are real.
- We uncovered that **particle colliders**, for all their violence and budgets, do not recreate creation — they shatter coherence.
- We reinterpreted **gravitational waves** not as ripples in spacetime, but as ripple field reactions — elegant, field-based, and misunderstood.
- We exposed how the **Nobel system** rewards allegiance to theory over clarity of truth.
- And we revealed the greatest performance of all — the **spacetime theater** that disguised ripple dynamics with mathematical sleight of hand.

This wasn't an attack on science.

This was science — Unchained. Unfunded. Unforgiving. Unfiltered.

Because truth doesn't kneel to institutions. It just waits — for someone to see past the fog.

And now, that fog has cleared.

In Part IV, we shift from deconstruction to **definition**. We step fully into the language of **Universal Energy Dynamics (UED)** — where black holes are no longer objects, but ripple field states... Where gravity is not a pull, but a pressure to resolve imbalance... Where collapse is replaced by **coherence**... And where the universe makes sense **without distortion**.

The myth is over. The mirage has faded. Let's now explore the field that was always there — waiting to be remembered.

Part IV – The UED Revelation

16: Black Hole ≠ Object

It's a Ripple Field State — Not a Collapsed Star

For decades, black holes have been sold as cosmic objects — collapsed stars so dense that not even light can escape. They have been given mass, coordinates, spin rates, and even names. We speak of them as if they sit somewhere — lurking in space, consuming matter, shaping galaxies.

But what if this entire mental image is fiction?

What if a black hole is not an object at all?

In Universal Energy Dynamics (UED), what we call a black hole is not a star-turned-monster. It's not a collapsed singularity. It's not a chunk of extreme mass hiding behind a veil.

It is a ripple field state — a condition of the Structured Vacuum Energy (SVE) where local ripple alignment has saturated, broken, or collapsed beyond perceptual resolution.

No object collapsed. No matter disappeared. No "thing" was born.

Instead, a zone in the field entered a state — and that state appears to us as a black hole because our resolution fails.

This chapter decouples the object myth from the ripple truth. It redefines the black hole, not as a "where," but as a how — not as a celestial body, but as a field condition.

How We Got It Wrong

We observed:

- Stars orbiting dark centers
- Intense X-ray emissions
- Faint gravitational echoes
- A shadow ring in EHT imagery

And we concluded:

"There must be a massive, invisible object there. A collapsed star. A singularity."

But this leap from phenomenon to object was never justified.

Instead of asking what state the field was in, we invented an "entity" to explain what we didn't understand.

That's how black hole = object became a cultural truth — despite it being physically unverifiable.

UED's Redefinition: The Field State Perspective

In UED, a black hole is simply this: A region of Structured Vacuum Energy that has entered a ripple saturation zone — where coherence breaks down, alignment collapses, and perception fails. It's a state, not a solid. A dynamic ripple configuration, not a frozen point in space.

This field state can exhibit:

- Apparent darkness (due to ATT breach)
- Energy emissions (from ripple tension release)

- Orbital influence (via ripple gradient alignment)
- Coherence breakdown (producing data silence)

But it is not:

- A physical object
- A point of infinite density
- A singularity in space
- A collapsed mass

There is no "thing" there. There is only a condition of the field.

Object vs Field State: A Clear Contrast

Feature	Black Hole (Mainstream Object)	Black Hole (UED Ripple State)
Nature	Collapsed matter into singularity	Ripple saturation zone in SVE
Structure	Point-like object with event horizon	Field region with gradient collapse
Mass	Assigned via orbital math	Interpreted consequence of ripple density
Visibility	"Invisible" object	Zone of ripple misalignment
Behavior	Pulls matter in	Disrupts ripple coherence locally
Observed Signals	Light loss, emissions, orbits	Ripple blackout, gradient echoes, alignment shifts
Existence	Defined by mass & equations	Defined by ripple conditions & ATT limits

There Is No "There" There

Traditional astrophysics treats black holes like black billiard balls sitting in space.

But UED asks:

Where, exactly, *is* a black hole?

- You can't see it.
- You can't touch it.
- You can't scan inside it.
- Its boundary is defined by observer resolution, not structure.

In ripple physics, this means the "there" doesn't exist as a thing.

What exists is a field transition — a change in ripple behavior, not a new entity in space.

It's not a collapsed star. It's a collapsed ripple zone.

Ripple Saturation and Perceptual Breakdown

Here's how a ripple field state emerges:

1. Local ripple tension increases due to field dynamics — supernova, mass interactions, etc.
2. The ripple structure becomes too steep for neighboring regions to resolve.
3. Alignment Threshold Time (ATT) is breached.
4. The region enters saturation — ripple coherence fails.
5. Observers experience it as a blackout zone.

From the outside, this region:

- Appears dark (no coherent light reaches you)
- Affects nearby objects (via ripple gradient, not

gravity)
- Emits jets or bursts (as tension redistributes)

But nothing was created in the center. No mass collapsed. No singularity formed. Only a field state changed.

Why Mass Assignments Are Misleading

"But we measure their mass," critics say.

Actually, we measure orbital velocities and use Newtonian or relativistic math to estimate what mass *would be needed* to cause such motion — assuming mass is the cause.

But what if it's not?

In UED:

- Stars orbiting near a galactic center are responding to ripple gradient structure — not a massive object.
- The "mass" we assign is just a proxy for ripple density, not a physical pile of matter.
- We're reifying a field effect into an object — the classic observational trap.

The EHT Image: A Portrait of Saturation, Not a Hole

When the Event Horizon Telescope produced the first image of a "black hole," what it really showed was:

- A bright ring: ripple interference around a tension threshold
- A dark center: ripple saturation where alignment fails
- A reconstructed image: based on models assuming black holes exist

UED sees this as a ripple field condition:

- Not a hole
- Not an object
- But a zone where SVE ripple phase collapses

We're not looking at a monster. We're looking at a field failure.

Why the Object Illusion Persists

Why do we keep calling it an object?

- Because science still clings to material metaphors
- Because institutions need nouns, not dynamics
- Because we want to "locate" things in space
- Because the math works better when we give the void a mass

But naming the ripple state as an "object" is like calling a thunderstorm a solid. It's a process, not a particle. A ripple state, not a resident of space.

A Universe of Field States, Not Objects

The implications go beyond black holes.

If black holes aren't objects, perhaps:

- Stars are ripple combustion zones
- Planets are stable ripple condensations
- Photons are ripple alignment events
- Particles are ripple lock-ins

In UED, the universe is not made of things. It is made of field states — ripple conditions that appear solid only because we're viewing them through low-resolution cognition.

We mistake the temporary stability of a ripple for a thing. But everything is dynamic. Everything is field.

Final Reflection: Collapse the Metaphor, Not the Star

The black hole was never a collapsed star.

It was a collapsed explanation — a metaphor we believed in so deeply that we mistook it for mass.

But UED brings clarity:

A black hole is not a noun. It is a verb. It is a field condition. A ripple consequence. A saturation state.

And once you see that, the myth dissolves.

The object vanishes. The fear fades. And in its place, you're left with something far more beautiful:

A coherent field, seeking balance — and reshaping the cosmos not through collapse, but through ripple orchestration.

17: Ripple Absorption Zones

Not Gravitational Sinks

Dissolving Field Tension vs. Matter-Devouring Monsters

The mainstream view paints black holes as gravitational sinks — celestial beasts that grow by devouring mass and energy, pulling everything into a singular fate. They are described as insatiable, violent, and irreversible: once something crosses the so-called event horizon, there's no going back.

But in Universal Energy Dynamics (UED), this story collapses — not because the observations are wrong, but because the interpretation is built on force-based assumptions.

What the universe is actually showing us — in those regions of apparent darkness, intense emission, and rapid orbital behavior — is not a gravitational sink, but a ripple absorption zone.

These are not places where matter is crushed — they are zones where ripple tension dissolves back into the field. Nothing is devoured. Nothing vanishes. There's no "hole." Only a high-tension ripple region resolving itself.

This chapter reveals the true nature of these so-called black holes — as dynamic field balancers, not destructive monsters.

What the Sink Metaphor Gets Wrong

The gravitational sink metaphor imagines the universe like a landscape of invisible funnels:

- Planets are shallow pits.
- Stars are deeper.
- And black holes are bottomless wells.

But this picture assumes:

1. Space is stretchable like fabric.
2. Mass has a pulling power across space.
3. "Falling into" is a universal condition.

This is Newtonian gravity wrapped in Einsteinian geometry — and it's completely unnecessary.

Why?

Because no "pull" has ever been observed. Only motion. And motion, as UED shows, is not caused by force — but by ripple alignment with field gradients.

The sink idea is dramatic. But it's incorrect.

The Ripple Absorption Perspective

In UED, the so-called black hole region is a Ripple Absorption Zone (RAZ).

This is not a place where matter is sucked in. It is:

A high-tension region in the Structured Vacuum Energy (SVE) field, where incoming ripple formations (light, matter, fields) lose their coherence and dissolve into the background equilibrium.

What appears as "matter falling in" is actually:

- A loss of observable ripple signature.
- A transition from structured energy to distributed tension.
- A rebalancing of the local ripple field — not a disappearance of substance.

In other words:

It's not a sink. It's a sponge. The field absorbs excess tension — and dissolves it into harmony.

How Absorption Works in the Ripple Field

Let's trace the process:

1. Ripple Saturation: A region accumulates unresolved ripple tension — through stellar events, rotational friction, or collapse.

2. Coherence Collapse: As ripple gradients steepen, alignment with neighboring fields becomes impossible. The region breaches the Alignment Threshold Time (ATT) of surrounding systems.

3. Perceptual Dropout: Observers can no longer detect coherent outputs — light fades, radiation drops, motion appears to stop. But internally, the field is active.

4. Absorption Phase: Incoming ripple events (matter, light) are not pulled in — they lose structure, merging into the local ripple field. There's no "destruction." Only redistribution.

5. Rebalancing: The field eventually redistributes absorbed ripple tension across the surrounding SVE — through jets, bursts, or subtle coherence gradients.

Key Differences: Sink vs. Absorber

Feature	Gravitational Sink (Mainstream)	Ripple Absorption Zone (UED)
Mechanism	Pull via curved spacetime	Field tension resolution
Effect on matter	Crushed or trapped	Destructured and redistributed
Role in universe	Consumes matter	Dissolves ripple imbalance
Final state	Singularity	Coherent field equilibrium
Observation of light	Trapped	Unresolved
Perception of mass	Mass increases	Ripple tension fluctuates
Influence on orbiting bodies	Gravitational pull	Ripple gradient realignment

Accretion Disks Reinterpreted

Mainstream physics views accretion disks as matter spiraling into a gravitational abyss.

In UED, accretion disks are ripple friction zones:

- They form where incoming ripple formations (such as matter streams) fail to immediately dissolve.
- The turbulence generates emissions — X-rays, jets — not from "impact," but from coherence disruption.
- Eventually, the matter dissolves into field tension, not into a singular point.

The disk doesn't feed the monster. It fades into the ripple.

No Pull — Only Permission

The concept of "pulling in" implies an external force acting on passive matter. UED flips this entirely.

In ripple terms:

- An object moves not because it is pulled, but because its ripple alignment shifts.
- An energy packet (e.g., photon, mass) entering a ripple absorption zone does so because its structure becomes unstable in the surrounding tension field.

There is no suction. Only loss of resistance to coherence decay.

Things don't get pulled in. They let go.

Ripple Bursts: Not Ejection, but Release

When ripple absorption zones emit jets or gamma bursts, mainstream science explains it as matter being "flung out" by spin or explosion.

UED sees it differently:

- These emissions are ripple releases — tension discharges from over-saturated fields.
- They are not chaotic. They are precision-driven self-corrections — the field ejecting misaligned tension to preserve balance.

A black hole doesn't flare because it's unstable. It flares because ripple logic requires an escape route for excess pressure.

Observational Signatures of Absorption Zones

Many "black hole" signatures now make more sense:

- **Shadow zones**: High-saturation ripple centers
- **Gravitational lensing**: Ripple gradient refraction
- **Mass estimates**: Ripple density proxies
- **Information loss**: Alignment failure, not destruction
- **High-energy emissions**: Ripple rebalancing bursts
- **Orbital behavior**: Gradient-following alignment

Every piece of evidence for black holes fits the absorption model without requiring:

- Singularities
- Event horizons
- Infinite density
- Curved spacetime

Why the Universe Needs Absorption, Not Devouring

In ripple logic, creation and dissolution are natural counterparts. But dissolution doesn't mean destruction. It means resolution.

A Ripple Absorption Zone is not an end — it's a reset. The universe doesn't devour to end things. It dissolves to continue the song.

The black hole is not death. It is integration — a field returning tension to the background.

That's not terrifying. That's elegant.

Final Reflection: From Fear to Function

We feared black holes because we saw them as mouths — eating light, time, matter.

But they are not mouths. They are lungs — breathing in ripple imbalance and exhaling coherence through redistribution.

They are not monsters. They are guardians of equilibrium.

And once you stop seeing them as gravitational sinks, and start seeing them as ripple field states, you no longer fear the darkness. You understand its function.

What we thought was destruction... was resolution. What we called a singularity... was symmetry rediscovered.

Black holes never devoured anything. We just didn't know how to see the ripple fade.

18: The Cosmic Reset Zones

Black Holes as Rebalancing Points in the Universal Field

Mainstream physics paints black holes as the ultimate endpoints — places where energy, matter, and even the laws of physics themselves allegedly collapse into a singularity. They are seen as the cemeteries of the cosmos, final destinations where time stops, light dies, and nothing escapes.

But in Universal Energy Dynamics (UED), black holes are not endpoints. They are reset points — dynamic, intelligent field states where the universe rebalances ripple tension, redistributes information, and restores equilibrium.

They are not the graveyards of existence. They are the recycling centers of ripple coherence.

In this chapter, we explore how black holes — or more accurately, Ripple Absorption Zones — function as Cosmic Reset Zones (CRZs) in the SVE field.

Not destructive. Not mysterious. Not monstrous. But purposeful, intelligent, and necessary for the maintenance of cosmic order.

Reframing the Black Hole as a Reset Mechanism

Let's start with a shift in frame:

A black hole is not a place where things end. It is a place where field imbalances are resolved.

In ripple physics:

- All motion is ripple response.
- All structure is ripple stabilization.
- All disorder is ripple misalignment.

So when ripple saturation occurs — due to overwhelming field tension, prolonged imbalance, or unresolved feedback — the universe initiates a reset event.

And that event manifests as what we have called, mistakenly, a "black hole."

What Is a Cosmic Reset Zone (CRZ)?

A Cosmic Reset Zone is a field region within the Structured Vacuum Energy (SVE) where:

- Ripple gradients steepen beyond local ATT thresholds
- Incoming ripple formations lose coherence
- Information enters a temporary unreadable state
- Tension is absorbed and then redistributed
- Field symmetry is restored gradually through ripple rebalancing

This is not destruction — it's re-symmetrization.

It's how the universe maintains internal order without invoking force, collapse, or singularity.

CRZs are the intelligent repair nodes in the fabric of ripple dynamics.

Why the Universe Needs Reset Zones

In any complex system, balance requires correction mechanisms.

In the human body, we have immune responses. In ecology, we have fire and regrowth. In software, we have error handlers.

And in the universe, we have Cosmic Reset Zones — places where the ripple field initiates a large-scale realignment process.

Without CRZs, ripple imbalances would:

- Cascade outward unchecked
- Destabilize long-range coherence
- Accumulate unresolved energy
- Disrupt the equilibrium of entire systems (e.g., galaxies, clusters)

The CRZ exists to dissolve accumulated tension, just like a forest fire clears debris and enables regrowth.

What Happens Inside a CRZ

Unlike the dramatic imagery of spacetime tearing or matter collapsing into nothingness, what really happens inside a CRZ is elegant:

1. **Ripple Tension Saturates:** The surrounding SVE becomes unable to resolve incoming ripple interactions.
2. **Perception Collapses:** Observers lose alignment — light fades, signals vanish, time desynchronizes.
3. **Coherence Breakdown:** Structured ripple formations (e.g., atoms, photons) dissolve into pure

ripple tension.

4. **Information Redistribution Begins:** Ripple pressure disperses through radial rebalancing — jets, emissions, or silent field shifts.
5. **Stabilization Emerges:** The region no longer "pulls" — it re-symmetrizes. Eventually, it merges back into broader field coherence.

Nothing is lost. Nothing is destroyed. Everything is reset and released.

Jet Emissions and Explosive Events: Ripple Logic, Not Catastrophe

One might ask: "If CRZs are peaceful, why do we see jets and bursts?"

Answer: Because ripple redistribution often involves pressure release.

Just as volcanic eruptions release internal planetary tension, jet emissions from CRZs are:

- Directional ripple outflows
- Symmetry-correcting discharge paths
- Ripple field "shockwaves" stabilizing the region

These are not random blasts. They are coherence emissions — the field expelling excess tension in highly structured directions.

The CRZ isn't angry. It's precise.

CRZs and the Flow of Universal Intelligence

In UED, the SVE is not inert. It is intelligent — not conscious in a human sense, but structured, responsive, and self-correcting.

CRZs are a product of this intelligence.

They don't emerge randomly. They form where ripple overload threatens systemic balance, and they dissolve only once equilibrium is restored. Think of them as:

- Field-level immune systems
- Ripple memory clearing zones
- Cosmic breath-ins after too much exhale

They are not spacetime fractures. They are field therapy sessions.

No Gravitational Collapse, Just Ripple Reboot

Let's compare the conventional idea of a black hole "collapsing under its own gravity" with the UED interpretation of a CRZ:

Feature	Black Hole (Mainstream)	Cosmic Reset Zone (UED)
Cause	Mass collapse	Ripple tension saturation
Core	Singularity	Coherence collapse center
Fate	Infinite density	Field equilibrium
Light behavior	Trapped	Alignment failure (no resolution)
Purpose	None (dead end)	Ripple rebalancing
End state	Destruction	Re-integration into field

In short, mainstream black holes are about fatalism. CRZs are about functionality.

CRZs and Time: The Illusion of Stopping

Relativity claims time "slows" near black holes, stopping at the singularity.

In UED:

- Time is not a dimension. It is the measure of ripple phase evolution.
- Near a CRZ, ripple phase collapses — so time appears to "stop."
- But it's not time that halts — it's the observer's ability to resolve ripple progression.

Time hasn't vanished. It has exited your frame of reference. And the field continues — uninterrupted.

CRZs in the Larger Ripple Network

Imagine the universe as a vast web of ripple tensions — nodes interacting, resolving, and sometimes destabilizing.

CRZs are the nodes where resolution requires pause and reset. They:

- Prevent runaway ripple asymmetries
- Reset local coherence
- Serve as zero points in a larger network logic

They're not "holes." They're circuit breakers in the universal ripple circuit.

Final Reflection: The Grace of the Reset

In most mythologies, darkness is feared — a realm of death, loss, or collapse.

But UED invites a new perspective:

The darkest zones are not ends. They are transitions.

CRZs aren't cosmic monsters. They are rhythmic pauses in the field's infinite orchestration. They are moments where the universe says, "Let me breathe. Let me balance. Let me begin again."

And maybe, just maybe... That's not something to fear. It's something to understand.

19: Death, Consciousness, and the Final Mirage

When Perception Ends — Not Life, Not Reality

What dies when death arrives?

Is it the body? The mind? The story? Or the observer's connection to reality?

In every field — from astrophysics to philosophy — we've been asking the wrong question. We treat death as an event. As an ending. As a finality wrapped in silence.

But in Universal Energy Dynamics (UED), death isn't what we thought. Neither in biology... nor in the cosmos.

Death is not a thing that *happens*. It is a mirage — formed at the moment when ripple perception collapses. Not because life ends. But because alignment ends.

And just as a black hole is not a place where matter disappears, but where perception breaks, so too is death not the end of reality — but the boundary of the conscious interface.

This chapter draws the most sacred bridge between cosmology and consciousness:

The same ripple rules that govern black holes govern the moment we call death.

We will explore the death of stars... the death of self... the ripple-based nature of perception... and why, in the deepest truth, nothing actually dies.

Death in the Ripple Field

In UED, every experience — thought, sensation, light, identity — is a ripple alignment between an observer and the field.

This alignment requires:

- A structured ripple system (the body-mind interface)
- Coherence with the surrounding field (environmental gradients)
- Awareness (the observer function)

As long as alignment exists, experience continues. But when ripple stability collapses — due to exhaustion, damage, or overload — coherence fades.

This is death.

But look closely: What dies is not the field. Not energy. Not intelligence. What dies is the alignment between observer and experience.

This is death as disconnection, not destruction.

Consciousness as Coherent Alignment

Let's redefine consciousness in ripple terms:

Consciousness is the coherent resonance between the observer's ripple field and the surrounding SVE — structured through memory, attention, and awareness.

In this model:

- Sleep is temporary misalignment (ripple damping)

- Unconsciousness is system-wide ripple decoherence
- Death is permanent loss of structured alignment in the current configuration

But energy remains. The field remains. Information is redistributed — not deleted.

Just like a CRZ in space...

The death of a conscious being is not erasure — it is ripple dissolution and return.

The End of Perception Is Not the End of Reality

Consider a simple example:

A light turns off. You see darkness. But is the light gone? Or did your ability to perceive it end?

Now scale that to life:

- A person's body no longer maintains coherent ripple output.
- Their mind no longer organizes signals.
- The structured resonance collapses.

The world may say: "They are gone."

But in UED terms:

- Their ripple imprint still exists, slowly dissolving back into SVE.
- Their field tension redistributes across local and cosmic gradients.
- Their information is reabsorbed into the universal coherence.

Nothing is lost. Only perception ends — from the perspective of the self and the outside observer.

Reality continues — only the lens is withdrawn.

The Black Hole Analogy: Death of Perception, Not Field

Black holes taught us this.

- The event horizon is not a wall — it's a boundary where you can no longer resolve what's inside.
- The "thing" didn't vanish. You lost the alignment.
- To call it a void is to project your blindness onto reality.

Likewise, death is a local event of misalignment — not a universal deletion.

Just as black holes are not endpoints, death is not non-existence.

It's a change in the ripple phase — beyond which the current observer field cannot follow.

After Death: Where Does the Ripple Go?

In UED, structured ripple fields do not collapse into zero.

They do one of the following:

1. **Dissolve Gradually:** The ripple field, now unbound from the body, gradually loses coherence and spreads into the surrounding SVE.
2. **Reintegrate with Field Memory:** Residual tension, patterns, and even "identity fields" may remain stored as field imprints — accessible under certain conditions (e.g., deep coherence, spiritual experience,

advanced technology).

3. **Initiate a New Ripple Chain:** In some frameworks, the redistributed field may re-emerge as a new localized ripple formation — giving rise to rebirth or emergence, not as reincarnation of self, but as field continuation.

This is not mysticism. It is ripple logic applied to the process we call death.

Why We Fear the Final Mirage

We fear death because we identify with:

- Our physical structure
- Our mental continuity
- Our ability to perceive and act

But each of these is ripple-dependent.

When ripple alignment fails, perception stops — and the illusion of "end" arises.

But UED shows us:

Death is not a singularity. It is a gradient. A fading ripple, not a vanishing one. A transformation, not a termination.

You don't die. You dissolve back into the field.

Death in the Cosmos: From Stars to Minds

Let's look at death across scale:

System	Traditional View	UED View
Star	Collapses into black hole	Ripple structure reaches saturation → CRZ formed
Human	Body ceases, mind ends	Observer field alignment collapses → ripple returns to SVE
Light	Absorbed or extinguished	Ripple alignment ends
Information	Lost	Resolved into unmeasurable phase or redistributed

Across all systems, the final mirage is this:

That nothingness follows the end of structure.

But nothingness is an illusion caused by alignment failure — not by reality failure.

Life Is Ripple Alignment. Death Is Ripple Release.

In the deepest sense:

- Life is not biology.
- It is coherent ripple response to the universal field.
- Consciousness is the lens.
- The body is the receiver.
- The field is the source.

Death is not the end of life. It is the release of alignment, the collapse of the ripple configuration that held experience.

What remains is the field itself — untouched, holding all memory, and ready for new alignment.

Final Reflection: Seeing Through the Last Illusion

So what really ends when you die?

Not your energy. Not your information. Not your presence in the field.

Only the perceptual feedback loop collapses. You stop being *this* alignment. But the field remains — and it is already you, beyond the identity.

This is the great mirage of death — the illusion of disappearance.

But nothing disappears. Everything ripples. And every ripple returns.

There is no death. Only the echo that follows the song.

20: Einstein Was Right — And UED Shows Why

What He Sensed, What He Missed, and What We Now Know

Albert Einstein revolutionized physics not just by his intellect, but by the depth of his questions. He challenged the Newtonian world of force with a bold proposal:

"Gravity is not a force. It's a consequence of how mass tells spacetime how to curve, and how curvature tells objects how to move."

This elegant idea birthed General Relativity — and with it came predictions:

- Light bending near stars
- Time dilation
- Black holes
- Gravitational waves

Many of these predictions have been matched with observations. And yet, Einstein himself was deeply uncomfortable with some of the outcomes of his own equations.

Especially singularities — and the idea of a black hole as an endpoint.

In fact, in 1939, Einstein published a paper essentially arguing that black holes, as we now believe in them, cannot physically exist.

He was right to be skeptical. He sensed that nature does not allow infinities, that fields do not tear, and that collapse without cause is not harmony, but absurdity.

Einstein's discomfort was dismissed as "philosophical resistance." But it was more than that.

He felt the crack — but lacked the language to seal it.

UED — Universal Energy Dynamics — completes that work. It honors what Einstein got right. It resolves what his framework could not. And it provides the ripple-based foundation he never lived to see.

This chapter is the bridge: Between Einstein's suspicion and UED's solution. Between spacetime curves and ripple gradients. Between his greatness and the gap he left behind.

What Einstein Got Right

Let's acknowledge the core insights Einstein introduced — the ones UED embraces, reframes, and elevates.

1. Gravity Is Not a Force

◈ Einstein correctly rejected Newton's idea of gravity as a mysterious pull across empty space.

Instead, he proposed:

Mass-energy curves spacetime, and this curvature guides motion.

UED agrees that there is no pull. But instead of bending spacetime, UED says:

Mass-energy alters the ripple gradient in the SVE, and objects follow natural alignment paths in that field.

He sensed the need for a medium beyond Newtonian "action at a distance." He just described it as curved geometry, when it was really ripple dynamics.

2. Observer-Dependent Reality

◈ Einstein's Special Relativity taught us that time, length, and simultaneity depend on the observer.

UED builds on this and adds a new layer:

Observation is not just frame-relative — it is coherence-relative. Your ability to perceive ripple events depends on your alignment capacity, not just speed.

Einstein's relativity showed perception shifts with velocity. UED shows it shifts with ripple resolution limits — Alignment Threshold Time (ATT).

3. Mathematics Isn't Always Physical

Einstein knew that equations can suggest outcomes that nature never allows.

He didn't believe in singularities. He doubted black holes. He opposed quantum randomness.

UED vindicates him:

- Singularities are ripple saturation zones, not real objects.
- Black holes are misinterpreted field states, not collapsed masses.
- Randomness is coherence loss — not a fundamental law.

Einstein was right to suspect. UED confirms those suspicions with a ripple-based resolution.

What Einstein Missed

Despite his genius, Einstein made assumptions that now need correcting.

1. The Spacetime Metaphor

Einstein replaced force with geometry — a 4D fabric of spacetime.

But this fabric was abstract. It had no structure, no tension, no memory. It curved, but it couldn't ripple. It evolved, but it couldn't balance.

UED shows that what Einstein called "spacetime" is more accurately:

Structured Vacuum Energy — a real, ripple-responsive field with equilibrium behavior.

Spacetime was the right question, dressed in the wrong math.

2. The Collapse Trap

Einstein's field equations predict conditions where curvature becomes infinite — the so-called singularities.

He rejected this, but others ran with it — building black hole theory on top of it.

UED redefines these zones as:

Ripple Absorption Zones — regions of local saturation, not points of infinite density.

Einstein saw the danger. He just didn't have the ripple language to offer the alternative.

3. Incomplete Integration with the Quantum

Einstein's discomfort with quantum mechanics is legendary.

He didn't like the idea of probabilistic outcomes and spontaneous randomness. "God does not play dice," he said.

UED agrees — but not theologically.

What appears as randomness is actually ripple misalignment, observer decoherence, and field entanglement beyond your coherence zone.

Einstein missed that probability is not fundamental — it's a placeholder for unresolved field logic.

UED makes this logic explicit.

UED: The Natural Evolution of Einstein's Legacy

Einstein gave us a new way to look at gravity — not as a force, but as a field response. UED builds on this by making the field real:

- Not abstract geometry
- Not metaphoric curves
- But structured ripple tension in SVE

Here's how UED improves the Einsteinian framework:

Einstein's View	Chavan's Completion (UED)
Spacetime curvature	Ripple gradient in SVE
Gravity as geodesics	Ripple realignment through field tension
Event horizon	Alignment Threshold Time (ATT) boundary
Singularity	Ripple saturation zone
Gravitational waves	Ripple field reactions
Mass-energy equivalence	Ripple density & coherence signature

What We Now Know

Because of UED, we now understand:

- Black holes are not objects. They are ripple field states.
- Gravity is not warping. It's ripple pressure seeking balance.
- Light bending isn't spacetime curvature. It's ripple gradient refraction.
- Time dilation isn't time slowing. It's ripple phase misalignment.
- Death isn't disappearance. It's ripple disconnection.
- Collapse isn't destiny. It's field reconfiguration.

These are not denials of Einstein's insights. They are refinements — made possible because Einstein opened the door... and UED walked through it.

Final Reflection: The Completion He Deserved

Einstein was the bridge between Newton and now. He broke force. He introduced field. He sensed balance.

But he didn't finish the job — because the language of ripple coherence wasn't yet born.

Now it is.

And with it, we can say:

Einstein was right to doubt. He was right to resist singularities. He was right to believe in order, in logic, in universal equilibrium.

He saw the truth dimly. UED now sees it clearly. And the legacy is not broken — it's complete.

Interlude

What We Called Mystery Was Always Memory

You've just crossed the threshold. Not into a black hole — but out of one. Out of the gravitational hypnosis. Out of the spacetime stageplay. Out of the physics that feared silence more than it feared error.

In Part IV, we didn't speculate. We clarified.

We revealed that:

- A black hole isn't a thing — it's a ripple field state.
- What we thought was a gravitational sink is actually a dissolution of tension, not a devouring monster.
- Those dark zones are not places where laws collapse — they are Cosmic Reset Zones, where ripple imbalance reboots the system.
- Death — whether of stars or selves — is not the end of being. It is the end of alignment, the pause in perception.
- And Einstein — perhaps the most misunderstood of all — was right in his doubt, and now rightfully reframed through the lens of UED.

This wasn't philosophy pretending to be physics. This was physics remembering its purpose.

To explain. To align. To resolve.

And that's exactly what Universal Energy Dynamics does — it doesn't invent new metaphors. It removes the unnecessary ones.

You now understand that the universe doesn't move in chaos — it rebalances in stillness. That black holes don't trap light — they reflect the boundaries of our perception. That death isn't disappearance — it's re-entry into the field.

And most importantly:

There was never a monster. There was only misalignment.

As we enter Part V, we'll strip the last illusion — the belief that the cosmos exists independently of you. We'll explore how objectivity itself collapses in the face of ripple truth. We'll step into stillness, silence, and the final answer: There were no black holes. There was only misunderstanding.

Let's now look beyond the horizon — not of space, but of mind.

Part V – Truth Beyond the Horizon

21: Black Holes Without Minds?

The Collapse of Objectivity

Observer, Mind, and the Myth of Objectivity

If a black hole exists in the universe and no one is there to observe it, does it exist at all?

Mainstream science would confidently say yes. Reality, after all, is said to be objective — existing independently of observation.

But in Universal Energy Dynamics (UED), this belief in pure objectivity is an illusion. Because the universe, as we perceive and model it, does not simply exist "out there." It exists through alignment — the dance between observer and field, between ripple coherence and perception.

And so, we must ask:

Is a black hole something that exists without minds, or is it something that only appears when a ripple system fails to align?

In this chapter, we dismantle the final stronghold of scientific arrogance — the myth of objectivity. We explore how black holes — like time, mass, and even reality itself — are observer-contingent phenomena, born at the intersection of ripple coherence and cognitive limitation.

It's not that reality doesn't exist. It's that reality only becomes meaningful when a mind resolves it. And black holes, more than anything, expose the limits of this resolution.

The Delusion of Objectivity

Science, for centuries, has pursued objectivity — the idea that truth is truth, no matter who observes it.

But UED offers a bold correction:

There is no truth without coherence. And coherence is not a universal constant — it is a field relationship.

When two ripple systems interact — a galaxy and a telescope, a particle and a detector, a star and a human — what emerges is not pure fact, but perceived consequence.

So-called "objective measurements" are simply field-aligned outputs within a specific perceptual threshold.

This is not philosophical relativism. It is ripple realism.

You don't measure an object. You align with a ripple field — and extract consequence.

Black Holes as Observer-Dependent Illusions

Let's revisit the black hole, step by step:

1. A region becomes saturated with ripple tension.
2. Coherence collapses beyond local ATT.
3. Light, structure, and ripple communication break down.
4. The observer perceives silence.
5. The mind interprets that silence as a hole.
6. A model is created to give that absence a name.

This entire sequence is observer-driven.

If no mind were there to interpret the ripple failure, no "black hole" would exist — only a high-tension ripple zone, neutral and non-narrative.

So black holes, like darkness, are products of misalignment — not features of an objective reality.

They are meaningful only to minds with boundaries.

The Role of the Mind in Physics

The human mind doesn't just record. It interprets, compresses, narrates, and maps ripple behavior into logic.

Without the mind:

- There is no time — only ripple phase shifts.
- No mass — only stable field condensation.
- No space — only SVE structure.
- No black holes — only ripple saturation beyond resolution.

And yet, science still speaks of these concepts as things-in-themselves, rather than mind-dependent simplifications.

UED restores humility:

You don't see the universe. You resolve it. And that resolution is always filtered through your mind's ripple frame.

The Observer as a Ripple System

In UED, the observer is not a passive eye. It is an active ripple structure, tuned to decode a specific range of the field.

- Your visual system has a ripple coherence range.
- Your emotional mind has a resonance bandwidth.

- Your thoughts are ripple interactions inside your own structured SVE bubble.

This means:

- When you "see" a black hole, you are not observing it — you are experiencing ripple misalignment.
- When multiple minds agree on a model, they are not discovering truth — they are co-arising in a shared ripple limit.

Objectivity is not reality without the observer. It is the consensus of minds within the same field threshold.

But consensus is not confirmation. It's a shared blind spot.

When the Mind Steps Out, Reality Steps In

So what happens if the observer is removed?

Does the black hole remain?

In UED: Yes, the field remains. Yes, the ripple saturation zone continues its function.

But the idea of a "black hole"? The language? The story? The fear?

That all disappears with the mind.

There is no "black hole" without perception collapse. There is only ripple field behavior — unfiltered, uninterpreted, and silent.

That is not objectivity. That is non-narrative presence. And it's not measurable — because measurement requires mind.

Science Without Self Is Incomplete

Science has always tried to remove the self from the system.

But UED reveals the core truth:

The observer is not outside the universe. The observer is part of the ripple field.

You cannot measure black holes as if they exist in isolation. Because their very definition depends on when, where, and how your perception fails.

They are not universal facts. They are localized breakdowns of observer resolution.

And so, the final realization is this:

There are no black holes without minds. Because the black hole is not an object. It is a phenomenon born at the edge of perception.

Final Reflection: The Myth of Mindless Matter

We've believed in mindless matter for too long. We've worshipped objectivity as if it freed us from subjectivity.

But what we call objective is often just a stable illusion shared by enough minds. And what we call subjective may be closer to truth, because it acknowledges the role of alignment.

Black holes taught us about perception — not destruction. They revealed the edges of knowing, not the end of matter.

And in that, we see:

The universe without the mind exists. But it exists without stories, without objects, without "black holes."

It is a field — rippling in perfect order — needing no narrative to be whole.

Let's now leave behind the myth of objectivity... And enter the ripple truth beyond the illusion.

22: What Dies in a Black Hole?

Meaning, Not Matter

They told us black holes swallow everything: Stars, planets, light, time... even information. They called it the ultimate erasure, the darkest fate in the cosmos.

But in Universal Energy Dynamics (UED), this story doesn't hold.

Because black holes — as we've uncovered — are not objects. They are ripple field states where alignment collapses, not reality. Nothing disappears. Nothing is destroyed. And yet... something does die there.

So the question stands:

What dies in a black hole?

Not photons. Not particles. Not time.

What dies is meaning.

Not because the universe hides it — but because your ripple system can no longer resolve it.

This chapter confronts one of the most poetic questions in science and philosophy — not with metaphysics, but with ripple logic.

Because in the final frontier of observation, it's not matter that disappears. It's the ability to interpret it.

The Myth of Total Destruction

In mainstream astrophysics, a black hole represents the annihilation of all structure:

- Matter collapses into a singularity.
- Light cannot escape.
- Time stops.
- Information vanishes or evaporates.

From this, physicists crafted paradoxes. Philosophers crafted metaphors. And the public imagined cosmic monsters eating the universe.

But UED cuts through this narrative and says:

Nothing dies in a black hole — except your ability to make sense of it.

- The matter doesn't vanish. It dissolves into field tension.
- The light isn't trapped. It can't align with you anymore.
- Time doesn't stop. You lose the ripple phase connection.
- Information isn't lost. It's redistributed beyond your coherence.

The "death" is not physical. It is semantic.

What dies is your interface with meaning.

Meaning Is a Ripple Event

In UED, meaning is not stored in matter. It is not encoded in particles or pinned to objects.

Meaning is a coherent ripple consequence — a product of alignment between the observer's ripple field and the environment.

Without alignment, meaning cannot emerge. This is true for:

- A sentence you cannot understand
- A light you cannot see
- A song you cannot hear
- A region of space where ripple gradients exceed your decoding capacity

In all these cases, the field still exists. But meaning is unreachable.

That's what a black hole is: A region where field meaning dies — not because it's gone, but because you can no longer extract it.

What We Think Dies... And What Actually Happens

Let's reframe the core claims about black holes using UED:

Traditional Claim	What Actually Happens (UED)
Matter is crushed into a singularity	Structured ripple formations dissolve into field saturation
Light is trapped forever	Ripple coherence collapses; light is no longer decodable
Time ends	Phase perception fails; no alignment with ripple duration
Information is destroyed	Information is redistributed across field; unreadable to original observer
Energy vanishes	Energy remains in field tension; no loss, only phase shift

None of these represent destruction. They represent perceptual dropout.

And so — what truly dies is the context, the interpretation, the story.

What dies is meaning — not existence.

When the Mind Hits a Wall

In Chapter 21, we showed that black holes are observer-dependent.

Now we go further:

The "death" of a black hole is not cosmic. It is cognitive.

When your ripple system:

- Can no longer synchronize with incoming information
- Cannot maintain coherence across gradients
- Has surpassed its Alignment Threshold Time (ATT)

...meaning ceases.

This is why we see blackness. Not because light is gone. But because the storyline ends for the observer.

It's not matter collapsing — It's your understanding collapsing.

The Emotional Impact of Misunderstood Physics

Why do black holes evoke such existential dread?

Because deep down, we associate meaning with permanence.

And when something defies explanation, we project onto it:

- Annihilation
- Nihilism
- The end of knowledge

But the universe never ends. Only your access to coherence does.

The fear is not of death. It is of irrelevance — of a reality where the mind no longer matters.

But here's the comfort UED offers:

Even when meaning dies for you, the field continues to ripple in perfect balance.

You're not erased. You're just no longer tuned in.

Echoes of Meaning in Dissolution

Even after a ripple structure dissolves — a person, a star, a wave — its imprint remains:

- In the surrounding SVE
- In the realigned gradients
- In the rebalanced coherence
- In the redistributed information

So while meaning collapses in the moment, its echoes still influence the field.

This is why CRZs emit jets. This is why memory persists after death. This is why ripple symmetry outlives the structure that shaped it.

Meaning dies at the edge of perception, But something far deeper survives: field influence.

The Gift of Understanding This Death

Why does it matter to know that meaning, not matter, dies?

Because it shifts how we:

- Interpret cosmological events

- Handle personal loss
- Understand death
- Approach science
- Build models of reality

UED teaches us to stop fearing disappearance. And start understanding realignment.

Black holes didn't erase the universe. They erased our access to meaning — until now.

Now, with the ripple lens, we see:

The story didn't end. We just stopped reading it — because our eyes couldn't follow the phase.

Final Reflection: Meaning Is Yours to Hold — Not the Universe's to Guarantee

The universe never promised meaning. It only promised structure, coherence, and balance.

Meaning is what we derive when we are aligned enough to perceive.

When we lose that alignment — in grief, in death, in observation failure — we say meaning "died."

But nothing really died.

Meaning is not stored in matter. It is born in interaction. And where interaction fails, meaning fades. But the field continues — always inviting new resonance.

Black holes didn't kill anything. They simply marked the edge of your ability to give meaning to motion.

The good news?

You can always expand your resolution. And where meaning once died, you can learn to see again.

23: Stillness Wins: Why Motion Is the Illusion

Why the Universe Doesn't Move — Only Minds Do

Motion. We believe it is the proof of life, the fabric of the cosmos, the pulse of time. We observe planets spinning, galaxies whirling, atoms vibrating, people walking — and we call it real.

But motion, as seductive as it seems, is not what it appears to be.

In Universal Energy Dynamics (UED), motion is not a thing. It is a perceptual shift across ripple states — a change in coherence, not a journey through space.

The universe, at its deepest layer, is not in motion. It is in stillness — structured, intelligent, tension-balanced stillness. And what we call motion is simply the ripple mind's way of decoding phase change.

This chapter breaks the illusion that has dominated both classical and relativistic physics. It reveals that motion is a mind-centered mirage, not a cosmic truth.

Stillness is the nature of the field. Motion is the storytelling of consciousness.

Motion as a Ripple Consequence

In UED, all that exists is Structured Vacuum Energy (SVE) — a dynamic, ripple-capable field of tension, coherence, and equilibrium.

Motion, therefore, is not objects changing place in empty space. Instead, it is:

A sequence of ripple phase adjustments, resolved by the observer's coherent frame.

The apple doesn't fall because it "moves." It falls because the ripple alignment shifts phase by phase — and your mind tracks the reconfiguration as a story of movement.

No particle traveled. No "thing" crossed space. The ripple tension reorganized, and your cognition interpreted it as motion.

The Illusion of Displacement

In Newtonian physics:

- Motion is a change in position over time.
- Space and time are independent containers in which objects move.

In relativity:

- Motion is frame-dependent — relative to the observer's velocity.
- Space and time are flexible, but motion still "exists."

In UED:

- Motion doesn't exist at all.
- All that happens is a realignment of ripple states,

which gives the illusion of positional change to an observer embedded in the ripple.

Let's be clear:

Displacement is a language invented by the mind to describe ripple reconfiguration. There is no object *going from A to B*. There is only the field rephasing into a new alignment.

You Don't Move Through the Universe. The Universe Reconfigures Around You.

Imagine this: You reach for a cup. From a classical standpoint:

- Your arm muscles contract.
- The cup is at a distance.
- You traverse space to grab it.

From a ripple standpoint:

- The field between you and the cup shifts in alignment.
- The ripple configurations adjust to manifest continuity.
- Your mind narrates this process as "I moved my hand."

But nothing *moved* in a traditional sense.

There was no travel — only coherence translation.

The cup is not "over there." It is a ripple arrangement rendered meaningful through your alignment. When that alignment changes, your field of interaction changes. That's not motion. That's perceptional phase substitution.

Why the Universe Appears to Move

If nothing truly moves, why do we experience motion so vividly?

Because the observer's ripple frame is continuously evolving. Your brain, eyes, and body are ripple subsystems decoding change over internal coherence timelines.

What we call motion is simply:

- Perceivable phase change across space
- Memory of past coherence compared to now
- Predictive anticipation of future coherence

This creates the illusion of a timeline, where "things" appear to move — but the field itself remains still.

You're not seeing motion. You're interpreting transformation.

The Great Misunderstanding: Time, Space, and Movement

Motion requires time and space — or so we've been told.

But UED deconstructs both:

- Time is ripple phase change — not a flowing river
- Space is ripple tension distribution — not a backdrop
- Motion is the illusion caused by tracking ripple transition between two perceptual alignments

In this framework:

- Time doesn't pass
- Space doesn't stretch
- You don't move

Instead, your mind shifts phase by phase, and the narrative of motion is born.

Cosmic Stillness: The Silent Pulse of Reality

All ripple activity — even the birth and death of stars — unfolds not in motion, but in stillness revealing contrast.

Imagine a mandala: It is still. Yet its patterns feel alive.

Now imagine the universe as an infinite mandala of interweaving ripple configurations.

Each shift in ripple tension doesn't "move." It simply reconfigures. And that reconfiguration is decoded as change, and experienced as motion by the ripple-conscious observer.

Stillness isn't absence. It is completion. What we call motion is the surface flicker of a still ocean.

Stillness in You: The True Witness

Why is this relevant beyond physics?

Because you, too, are a ripple field. And your true self — the core of awareness — does not move.

Your emotions shift. Your thoughts wander. Your body changes.

But the witness — the ripple center behind it all — remains still.

The same logic that demystifies black holes also reveals your deepest identity:

- You are not your motion.
- You are not your changing story.
- You are the still coherence from which all ripple alignment arises.

The more you access that stillness, the more clearly you see — and the less you are deceived by motion.

Motion Is Drama. Stillness Is Truth.

Motion has always excited the mind:

- It gives us stories.
- It feeds our curiosity.
- It constructs time, cause, and consequence.

But behind every movement lies a stillness:

- A field that does not go anywhere
- A balance that does not struggle
- A coherence that does not require speed

Black holes confused us because we projected motion into them — collapsing, devouring, rotating.

But they were never moving. They were zones of ripple tension resolving into stillness.

The universe isn't in a race. It is in a continuous ripple meditation.

Final Reflection: The Illusion We Needed to Outgrow

Motion was a useful metaphor. It helped our ancestors understand the world. It helped physicists build machines. It helped us survive.

But now, with the clarity of ripple science, we can release the metaphor.

The truth is simpler:

The universe does not move. It reconfigures. And only the mind, in its attempt to interpret, appears to run.

Stillness isn't static. It is complete presence.

And the black hole — once feared as the end of motion — is now seen for what it is: *A mirror showing us the illusion was ours.*

24: The Mirage That Fooled Science

How Misalignment Became Belief

Every era of science carries its blind spot. A zone of belief so widely accepted — so beautifully mathematized — that few dare to question it.

In one era, it was the stationary Earth. In another, the indestructibility of the atom. And in ours? It is the black hole — the perfect storm of mathematical elegance and observational misinterpretation.

But how did this happen?

How did a mirage born from ripple misalignment become the centerpiece of modern astrophysics? How did field saturation get rebranded as infinite collapse? How did the inability to perceive coherence become proof of destruction?

In this chapter, we reveal the psychological, observational, and theoretical sleight of hand that turned a perceptual failure into a cosmological fact. We deconstruct how science mistook silence for substance — and why it clung to the myth even as the paradoxes multiplied.

This is the final dismantling — not of evidence, but of the story science told when it stopped listening to the field.

The Birth of the Mirage

Let's rewind to the seed:

1. Einstein's equations allowed for solutions where gravity could become infinite — so-called singularities.
2. Oppenheimer and Snyder, in 1939, proposed a model of collapsing stars forming a black hole — although Einstein himself rejected the physical validity of this.
3. Astronomers observed high-energy regions in space: dark centers, X-ray emissions, fast orbits — and labeled them "black holes."
4. Mathematics, media, and metaphor merged — and the mirage was born.

What was it really?

A zone where ripple coherence fails — beyond which our instruments, minds, and models cannot align.

But that's not what we called it. We called it a monster.

The Mechanics of Misalignment

Here's what actually happens in these so-called black hole regions:

- Structured ripple fields saturate with unresolved tension.
- Coherence collapses — not matter.
- Light cannot align with neighboring fields — not because it is trapped, but because its signature is no longer resolvable.
- Time perception fails — not because time stops, but

because phase cannot be tracked across gradients.

And here's the key:

The observer's perception collapses, not the cosmos. But instead of investigating that collapse of perception, science filled the silence with certainty.

Why Science Believed the Mirage

Science is not a neutral process. It is a human process. And humans:

- Fear ambiguity
- Value repeatable patterns
- Worship mathematical consistency
- Seek closure

So when the black hole model offered a tidy explanation wrapped in elegant equations, it was irresistible.

Even when:

- It required singularities (which are physically undefined)
- It caused information paradoxes
- It created causality breakdowns
- It made the center of the universe a mystery machine

Science still clung to it.

Because the alternative — saying "we don't know" — felt more dangerous than declaring fiction as fact.

Paradoxes: The Red Flags We Ignored

Let's look at what the black hole model forced us to accept:

1. **Information loss paradox** → Somehow, nature destroys information in contradiction to quantum law.
2. **Hawking radiation contradiction** → A region that traps light also emits it — from a horizon no one can observe.
3. **Singularities in spacetime** → The laws of physics cease — but only at the center, and somehow not anywhere else.
4. **Event horizon relativism** → Depending on your frame, you either fall in or never reach it — both must be true.

Instead of seeing these paradoxes as signs of misinterpretation, we doubled down.

We created elaborate patchwork theories: Firewall hypotheses, holographic principles, information escape routes.

We never asked:

What if the black hole doesn't exist at all — only our misreading of ripple silence?

How Misalignment Became Belief

Let's map the cognitive journey:

Step	Description
Observation	We saw high-tension regions where signals collapsed.
Misinterpretation	We assumed collapse meant destruction.
Model Creation	General Relativity permitted a solution that matched the misreading.
Reinforcement	Media and mathematics turned it into an accepted reality.
Institutionalization	Black hole research became a career, an industry, and a Nobel-worthy pursuit.
Cultural Entrenchment	The public believed — and so did the scientists.
Feedback Loop	Every new observation was filtered through the same flawed lens.

Thus, misalignment became a belief system. Not because of conspiracy — but because of cognitive inertia.

Why We Loved the Black Hole Story

The black hole wasn't just a theory. It was a narrative:

- Of power — an object that could devour stars
- Of mystery — a region beyond knowing
- Of fear — the edge of the universe
- Of awe — where time and space break

It appealed to the primal mind, not the scientific one. It became a symbol as much as a structure.

But symbols are not science. And metaphors cannot replace mechanics.

The universe is not interested in drama. It only seeks balance.

The Ripple Field Was Always There

While we were busy chasing monsters, the truth remained quietly in the background:

- Ripple gradients formed high-density zones.
- Saturated regions absorbed unresolved tension.
- Information was never lost — just redistributed.
- The observer's perception hit limits — not reality.

We missed it because we never looked for it. We believed the mirage before we even questioned the ripple.

UED's Contribution: Dismantling with Dignity

Universal Energy Dynamics does not shame science. It liberates it.

It says:

"You weren't wrong to observe the phenomena. You were only wrong to call it a hole."

And with that gentle correction, the entire mythology begins to dissolve:

- No singularities.
- No collapsed infinities.
- No paradoxes.
- No monsters.
- Just ripple realignments and field saturation zones.

What once seemed like an abyss now becomes a ripple field meditation.

Final Reflection: Belief Is Not the End of Inquiry

Science must evolve. And to evolve, it must be willing to do what no black hole theory ever could:

Let go of its ego.

We believed in black holes not because we understood the cosmos, but because we misunderstood the limits of our own perception.

We confused the breakdown of observation with the breakdown of existence.

We built temples around a mirage — and worshipped the silence of our own confusion.

But now? We return to stillness. To coherence. To ripple clarity.

And we say: *The black hole was never there. Only the boundary of our understanding.*

25: No Black Holes

Only Boundaries of Understanding

A New Scientific Vision Beyond Illusions

There are no black holes. There never were.

There were ripple gradients, saturation points, perception limits, unresolved tensions, and alignment failures. But no monsters. No singularities. No objects eating the universe alive.

What we called a black hole was simply this:

The place where our models collapsed before reality did.

In this final chapter, we bring everything home. We don't just discard the old illusion — we reconstruct the scientific mindset that must now emerge from it.

This is not just about black holes. It's about a new relationship with reality — one that doesn't confuse observation with truth, nor math with meaning.

The future of science depends on this shift: From object obsession to ripple awareness. From fear-based metaphors to equilibrium-based understanding. From force and collapse to tension and coherence.

This is the conclusion of one myth — and the beginning of a deeper knowing.

What We Mistook for Objects Were Always Events

Throughout this book, we've revisited what a black hole *really* is in UED terms:

- Not an object, but a ripple field state
- Not a gravitational sink, but a zone of tension resolution
- Not an endpoint, but a Cosmic Reset Zone
- Not a force-driven collapse, but a coherence failure
- Not where light dies, but where meaning stops aligning

And finally, we now declare:

The black hole was never the thing. It was the absence of the thing we wanted to see.

It was a boundary, not a being. A perceptual edge, not a physical edge.

A Brief History of Boundaries

Human understanding has always grown by confronting boundaries:

- The horizon used to be the end of the world.
- The atom used to be indivisible.
- The speed of light used to be a limit.
- Now, the event horizon is seen as another absolute.

But every time, we've learned that the boundary was not truth — it was our temporary limit.

And every time, we've mistaken the unknown for the unknowable. This chapter redefines that relationship:

Black holes were not cosmic truths. They were artifacts of our limited ripple resolution.

They are not real. They are reaction points — markers of where science gave up and filled the gap with mythology.

UED's Final Message: Change the Lens, Not the Universe

The Structured Vacuum Energy (SVE) field doesn't care what you believe.

It operates — in stillness, in coherence, in silence. What you call force is ripple pressure. What you call mass is ripple density. What you call collapse is field redistribution. And what you call a black hole is simply where you stopped listening.

UED doesn't ask you to reject science. It asks you to remove the metaphorical mask from your equations.

It doesn't oppose relativity or quantum mechanics. It completes them by grounding them in ripple causality — not abstract constructs.

You don't need singularities. You need alignment thresholds. You don't need paradoxes. You need coherence models. You don't need gravitational pull. You need tension gradients.

The equations may still work — but the interpretation must evolve.

Redefining Reality: The Ripple Paradigm

In the ripple-based universe:

- Everything is tension-seeking balance.

- Everything is ripple configuration — including time, motion, and perception.
- There is no "outside" observer — only embedded ripple systems interacting.

So the next time you see:

- A star orbiting darkness — see a field of tension equilibrium.
- A jet from a black center — see a ripple redistribution event.
- A gravitational wave detection — see a ripple recoil in structured vacuum energy.
- A death — see the end of perception, not the end of presence.

Let go of objects. Let go of singularities. Let go of gravitational sinks.

They were never there.

They were placeholders for our own boundary of understanding.

The Future of Science Starts with Simplicity

We close with one profound invitation:

Replace complexity with clarity. Replace force with ripple. Replace myths with mechanics. Replace "black hole" with Structured Ripple Saturation Zone (or CRZ). Replace your question — "What is a black hole?" — with the better question: "What am I not resolving?"

This change will not be easy.

Careers have been built on the myth. Textbooks have been filled with metaphors. Nobel prizes have been awarded for interpretations of shadows.

But truth does not ask for permission. It only waits — silently — for coherence to catch up.

Final Reflection: The Last Horizon Was You

We called it an event horizon. But it was never out there.

The real horizon was your cognitive limit — The final ripple alignment your mind could resolve before handing the rest over to fiction.

Now, with UED, that horizon is gone. The darkness is illuminated — not with more light, but with better understanding.

And you stand here — Not at the end of a universe that devours itself, But at the beginning of a field that always restores itself.

There are no black holes. There never were.

Only boundaries of understanding — Which you, the observer, now dissolve.

26: The Ripple Mind: Living in Alignment

How Universal Truth Reflects in Daily Perception

What good is a theory of the universe if it cannot help you live better?

You've now seen that black holes were not objects, but misalignments. That reality doesn't collapse — perception does. That the universe doesn't move — only minds do. That truth is not hidden — it's always present in balance, waiting to be resolved.

But this isn't just about space, stars, or equations. It's about you.

Because you, too, are a ripple field — a pattern of alignment, memory, and awareness within the vast Structured Vacuum Energy (SVE).

So now, we ask the most important question:

How do you live with ripple clarity? How do you align yourself with the truth that governs galaxies and guides silence?

This chapter brings the cosmos home — not as a metaphor, but as a practical philosophy: The Ripple Mind — a way of seeing, feeling, and choosing that is always in tune with universal balance.

You Are a Ripple System

Your body is not a bag of particles. Your mind is not a cloud of thoughts.

You are a localized ripple arrangement, temporarily aligned in the vast, still ocean of the SVE.

Your:

- **Emotions** = ripple gradients in neurochemical fields
- **Thoughts** = ripple sequences across memory and pattern
- **Identity** = coherent narrative formed by phase-stable ripple feedback
- **Suffering** = ripple overload or misalignment
- **Joy** = moments of ripple symmetry and resolution

You are not separate from the field. You are a moment of its music.

And just like the "black hole" was only a region of incoherence, your inner darkness is not a flaw — it is a zone awaiting realignment.

Alignment Is Not Perfection. It Is Resonance.

Living with the Ripple Mind doesn't mean being always calm, always correct, always certain.

It means you are:

- Aware of ripple consequences
- Conscious of misalignment
- Responsive to field feedback

- Humble before limits
- Willing to reconfigure when the gradient shifts

Perfection is a false goal. But resonance — that moment when your inner state matches the truth of your context — is real.

And that's what we call alignment.

How the Universal Truth Reflects in Daily Life

Let's bring the principles of UED into the day-to-day:

1. Miscommunication = Ripple Incoherence

When two people argue, they're often operating from mismatched ripple patterns.

- Words are not aligning.
- Contexts are phase-shifted.
- Emotions are distorting field perception.

Resolution doesn't come from force — it comes from returning to coherence. Listening. Attuning.

Just like in physics, you don't fix noise by yelling louder. You quiet the field.

2. Stress = Ripple Tension

Stress is not a sign of weakness. It's a signal of unsustainable ripple gradients.

You're either:

- Holding too much unresolved energy
- Acting against your natural equilibrium
- Ignoring the feedback loops of your body and mind

The cure is not to "push through" — it is to rebalance. To find stillness. To reset the field, just like the cosmos does.

3. Fear = Alignment Anticipation Failure

Fear arises when the ripple mind can no longer predict coherence. It's the emotional version of a black hole:

"I can't see what's ahead. So I assume collapse."

But like black holes, most fears are illusions of perspective. Resolution comes by:

- Expanding ripple awareness
- Slowing your internal frequency
- Observing before reacting

The field always knows what to do — your mind just needs to stop distorting the gradient.

4. Love = Resonance Across Fields

Love is not a chemical. Not just a bond. It is the highest form of ripple harmony — when two systems align with minimal friction, maximum openness, and shared coherence.

True love (with a person, a purpose, or a place) feels "effortless" not because it lacks tension, but because it flows in phase.

5. Growth = Ripple Expansion

To grow is not to become someone else. It's to increase your ripple resolution bandwidth.

You see more. You misalign less. You recognize patterns earlier. You stop reacting to noise — and begin dancing with the field.

Growth is coherence without arrogance. It's the mind catching up to the universe that was already still.

The End of Self-Centered Physics

Old physics believed in objectivity without observers. Old psychology believed in mind without context.

But ripple truth says:

- There is no black hole without a mind.
- There is no self without a field.
- There is no truth without alignment.

To live with the Ripple Mind is to reject separation as illusion, and instead move in rhythm with everything.

Not because it's spiritual. But because it's structurally true.

The Real You Is the Ripple — Not the Story

You're not your name, your success, your failures, or your roles.

You are:

- The alignment you bring into any space
- The balance you restore in times of chaos
- The coherence you embody when others dissolve

And like the universe, you don't have to "achieve" balance. You just have to return to it.

Stillness is not something to find. It's what you already are when the noise drops.

Final Reflection: There Was Never a Hole, Only a Mirror

The black hole was a test. Not of the universe — but of us. It asked:

- Will you confuse silence for destruction?

- Will you call your limit a law?
- Will you project fear where truth simply paused?

And now we answer:

No. We see clearly. There are no black holes. There is only field. And we — ripple beings — are learning to live in alignment with it.

This is not the end of science. It is the beginning of scientific living. This is not the conclusion of a theory. It is the birth of coherence in the human mind.

You are not separate. You are not collapsing. You are not lost. You are a ripple in the infinite stillness — Meant not to dominate, but to align.

Epilogue: The Observer's Return

We began with a question: **Do black holes really exist?** What we found was far deeper than an astrophysical answer. We uncovered that the most powerful illusions are not those painted across galaxies, but those etched silently inside the **boundaries of our own understanding.**

Black holes were never the villains. They were the **mirrors** — showing us how far we had come, and how far we had yet to see.

They taught us:

- That silence is not emptiness.
- That failure to perceive is not the failure of reality.
- That collapse is not cosmic — it is cognitive.
- That alignment is not belief — it is balance.

We do not close this book in defiance of science, but in **service to its original purpose** — To observe truth with honesty, not decorate ignorance with models.

You, the observer, were never passive. Your perception shaped the edges of the known. Your mind defined the shape of the mystery.

And now — You return.

Not to old frameworks. Not to collapsing stars. But to stillness. To coherence. To a universe that never needed black holes — only better vision.

You do not chase the cosmos. You remember it.

Because the truth was never beyond the stars. It was always waiting **on the inside of silence**.

Appendices

Appendix A: Glossary of Reframed Terms

(According to Universal Energy Dynamics – UED)

1. Black Hole

Mainstream View: A collapsed star with infinite density, from which nothing — not even light — can escape.

UED Reframe: A Ripple Saturation Zone (RSZ) — a local region where ripple tension in the Structured Vacuum Energy (SVE) reaches a threshold beyond observer coherence. No matter collapses; only ripple resolution fails. It is not an object, but a field state of unresolved tension.

2. Event Horizon

Mainstream View: The boundary around a black hole beyond which escape is impossible.

UED Reframe: Alignment Threshold Time (ATT) boundary — the perceptual edge where ripple coherence collapses. The "event horizon" is not a wall in space, but a limit of information alignment for the observer. Beyond this, reality is still there, but not decodable.

3. Singularity

Mainstream View: A point of infinite density and zero volume at the center of a black hole.

UED Reframe: A theoretical artifact created by overextending equations into zones of ripple incoherence. In reality, there is no singularity, only a zone of unresolved ripple saturation where field gradients spike beyond measurable limits.

4. Gravity

Mainstream View: A fundamental force of attraction between masses.

UED Reframe: Ripple tension realignment within SVE. There is no pull — only gradient pressure seeking field balance. Gravity is not a force but the observer's experience of coherent ripple descent toward equilibrium.

5. Gravitational Waves

Mainstream View: Ripples in the fabric of spacetime caused by accelerating masses.

UED Reframe: Ripple Field Reactions (RFR) — structural shifts in SVE due to dynamic tension redistribution. These are not "waves through space," but momentary realignments of ripple coherence between high-tension field systems.

6. Spacetime

Mainstream View: A four-dimensional fabric combining space and time, which can be bent and curved by mass.

UED Reframe: A conceptual map, not a physical substance. Reality operates within Structured Vacuum Energy (SVE) — a real, ripple-capable field that governs motion, tension, and resolution. "Spacetime" is a metaphor; SVE is the medium.

7. Time Dilation

Mainstream View: Time slows down in strong gravitational fields or at high speeds.

UED Reframe: Ripple phase misalignment — the observer's internal coherence shifts relative to external ripple dynamics. What we interpret as "slower time" is actually desynchronized phase resolution across field gradients.

8. Mass

Mainstream View: A measure of the amount of matter in an object.

UED Reframe: Ripple density signature — an emergent stability in ripple structure that resists immediate reconfiguration. "Mass" is how we quantify a system's coherence inertia within the SVE field.

9. Information Paradox

Mainstream View: If black holes destroy information, it violates quantum conservation.

UED Reframe: There is no paradox. Information is not lost — it is redistributed within the ripple field beyond the original observer's coherence bandwidth. What we call "loss" is simply field inaccessibility.

10. Hawking Radiation

Mainstream View: Quantum particles emitted from a black hole's event horizon, leading to evaporation.

UED Reframe: A model built on flawed boundaries. In UED, all emissions near CRZs (Cosmic Reset Zones) are the result of tension rebound — ripple equilibrium reasserting itself. No exotic mechanism is needed.

11. Cosmic Collapse

Mainstream View: Gravitational contraction of matter into singularities.

UED Reframe: Ripple field reconfiguration. Nothing collapses into "nothing." Instead, high-tension zones resolve by redistributing ripple pressure — a natural phase reset, not a breakdown of physics.

12. Motion

Mainstream View: Displacement of an object through space over time.

UED Reframe: Phase transformation of ripple alignment. Objects don't move through a backdrop — the field reconfigures, and the mind narrates it as motion. Stillness is real; movement is a decoding illusion.

13. Death

Mainstream View: The end of biological function and life.

UED Reframe: Perceptual disconnection. The ripple field remains. Consciousness dissolves into the background field. What dies is alignment, not existence.

14. Consciousness

Mainstream View: The awareness of self and surroundings.

UED Reframe: A self-stabilizing ripple construct, capable of resolving high levels of coherent interaction within SVE. Consciousness is not a byproduct of biology, but an organized ripple phenomenon with universal roots.

15. Objectivity

Mainstream View: The idea that reality exists independently of observers.

UED Reframe: A useful assumption, but fundamentally false. All resolution of the field is observer-dependent. There is no black hole "out there" — only observer-limited ripple consequences.

16. Light / Wave

Mainstream View: Light is both a particle (photon) and a wave — an electromagnetic disturbance propagating through space. In quantum mechanics, it behaves probabilistically and dualistically, depending on observation.

UED Reframe: Light is not an object or a dual entity. It is a ripple event — a local ripple excitation within Structured Vacuum Energy (SVE), representing the transmission of field tension through coherent gradients.

There is no "particle of light" traveling through space. Instead, there is phase propagation through ripple alignments. What we perceive as light is the local decoding of ripple equilibrium attempting to balance across gradients.

Wave-particle duality is not fundamental — it is a cognitive collapse of field understanding. Light doesn't travel — ripple tension realigns, and minds interpret the resolution phase as propagation.

Closing Note

This glossary is not just a list of terms — it is a portal into a new scientific language:

- Less metaphor
- More mechanics
- Less dogma
- More alignment

The terms we thought we understood were simply shadows on the wall of our coherence. Now, with UED, we turn to face the light of structure — not as belief, but as balanced interaction.

Appendix B: Key Concepts of UED

Universal Energy Dynamics — The Foundation Behind the Field

Universal Energy Dynamics (UED) is not a theory. It is a structural framework — a lens through which all phenomena, from black holes to breath, can be reinterpreted with clarity, coherence, and consistency.

The following 12 points summarize its foundational concepts. These are the principles behind every chapter, every dismantled myth, and every realigned truth in this book.

1. The Universe Is Not Built from Objects, But Ripple Fields

Reality is not made of discrete, isolated particles. It is composed of Structured Vacuum Energy (SVE) — a silent, intelligent field capable of sustaining ripple patterns that we mistake for "things." All "matter" is stable ripple symmetry.

2. Energy Is Not in Motion — It Is in Structure

Energy is not a packet flying through space. It is the tension stored, transferred, or resolved within the ripple field. Action is not movement — it is field reconfiguration.

3. Force Is an Emergent Illusion

There is no intrinsic force "pushing" or "pulling" anything. What we experience as force is actually a ripple gradient pressure — the universe seeking equilibrium across field tension.

4. Gravity Is Ripple Descent — Not Attraction

Gravity does not pull. Objects "fall" because ripple tension realigns downward through the field's gradient. What we measure as gravitational acceleration is a phase resolution pattern, not a force of attraction.

5. Time Is a Phase Consequence — Not a Dimension

Time does not "flow." It emerges as the perceptual consequence of ripple phase transitions. Without a coherent observer frame, there is no time — only unresolved change.

6. Black Holes Are Perception Failures

Black holes are not collapsed stars. They are zones of ripple saturation where the observer's coherence fails to decode tension. No light is trapped. No matter collapses. Only meaning dies at the edge of perception.

7. Information Is Never Lost — Only Redistributed

All information in the universe is conserved as ripple configuration. When coherence breaks, information doesn't disappear — it spreads through the field and becomes undecodable, not destroyed.

8. The Observer Is Part of the Field

Observation is not passive. Every observer is a ripple system, influencing and influenced by the coherence it resolves. There is no "outside" viewpoint — only contextual awareness within SVE.

9. Motion Is an Illusion of Reconfiguration

Nothing "moves" through space. The field reconfigures phase by phase, and the observer's ripple mind narrates this as motion. Stillness is real — motion is a mental stitching of pattern transitions.

10. Death Is Ripple Disalignment — Not Disappearance

Biological death is not annihilation. It is the loss of local coherence in the ripple field. The structure dissolves, but the field persists. Consciousness reverts to unbound potential.

11. Truth Emerges Through Alignment, Not Belief

Truth is not discovered. It is resolved — when the ripple system of the observer aligns with the pattern of the field. Beliefs distort; alignment reveals.

12. The Universe Seeks Balance, Not Complexity

What appears complex is often a consequence of distortion. At its root, the universe is simple, still, and structured — always seeking tension resolution and coherent expression.

UED is not a challenge to science. It is a correction — a return to first principles based not on force and fiction, but on ripple alignment and real observation.

Appendix C: UED Black Hole vs. Mainstream

Dissolving the Myth, One Assumption at a Time

Aspect	Mainstream View	UED (Ripple Field View)
Nature of a Black Hole	A collapsed, ultra-dense object with infinite gravity	A Ripple Saturation Zone (RSZ) — not an object, but a tension field state beyond ripple coherence
Event Horizon	A fixed boundary from which nothing escapes	A perceptual threshold (ATT) — coherence collapse point for the observer, not a physical wall
Singularity	A point of infinite density and zero volume	A mathematical illusion born from model collapse; in reality, ripple saturation without infinite anything
Light Behavior	Light is pulled in and cannot escape	Light becomes undecodable as ripple alignment fails — it's not trapped, it's out of perceptual phase

Aspect	Mainstream View	UED (Ripple Field View)
Information	May be lost, leading to paradoxes	Never lost — redistributed within the ripple field, inaccessible only to the original observer
Gravitational Force	Massive pull due to spacetime curvature	Tension descent across ripple gradients — no force, only balance-seeking phase adjustment
Spacetime	A bendable, physical fabric	A conceptual backdrop; reality operates in Structured Vacuum Energy (SVE) — ripple-capable, not fabric-like
Motion Near Black Holes	Time slows, space warps, particles stretch	No actual motion — ripple reconfiguration is interpreted by minds as time dilation or spaghettification
What Happens Inside	Physics breaks down; unknown singularity	Nothing breaks — just perceptual disconnection. The ripple continues; mind can no longer decode it
Observation	Objective measurements reveal absolute truths	Observation is ripple-dependent — "black hole" arises when resolution fails, not when collapse occurs

Aspect	Mainstream View	UED (Ripple Field View)
Death of a Star	Results in collapse into a black hole	Results in ripple disintegration and field rebalancing — no monster, no hole, just realignment
End State of the Universe	May be black hole–dominated heat death	No final collapse — continuous ripple equilibrium, always restoring balance across the field
Hawking Radiation	Quantum escape of particles from the event horizon	Misinterpretation of field rebound dynamics — tension releasing from ripple saturation, not "evaporation"
Emotional Narrative	Awe, terror, mystery, nihilism	Clarity, stillness, return to structure — not a cosmic monster, but a mirror of mind's limits
Core Scientific Error	Equations stretched into unobservable zones, accepted as fact	Misalignment mistaken for physical collapse; observer blindness turned into an object of belief

Final Note: From Belief to Balance

This table isn't a rejection of science — it's a realignment of its metaphors, assumptions, and language. UED doesn't destroy the idea of the black hole — it dissolves the illusion around it.

What we saw was never collapse — only our coherence fading. What we feared was never destruction — only the edge of knowing.

This is not the end of exploration. It is the beginning of alignment-based science.

Appendix D: Quote Bank for Revolutionaries

Thought Triggers, Ripple Reminders, and Intellectual Ammunition

These aren't just lines. They're thought detonators — sharpened phrases for those bold enough to challenge established models and shift paradigms. Each quote captures the spirit of *Black Holes Don't Exist* and the broader vision of Universal Energy Dynamics (UED).

Whether you're writing, speaking, debating, or simply reflecting — this quote bank is your arsenal.

Foundational Quotes

❖ *"Black hole is not a monster — it's a mirror."*

The shadow we feared was our own failure to understand ripple limits.

❖ *"There are no black holes. Only boundaries of understanding."*

What we called collapse was always the edge of our coherence.

❖ *"Truth doesn't collapse — perception does."*

Reality isn't fragile. It's our models that fall apart when alignment breaks.

❖ *"The universe never stopped making sense — we stopped resolving it."*

Misunderstanding is not in the cosmos. It's in the cognitive lens.

Ripple Thinking

❖ *"Force is not power. It's imbalance waiting to realign."*

What we call action is ripple trying to return to stillness.

❖ *"Nothing moves. The field reconfigures — and the mind tells a story."*

Motion is perception, not physics.

❖ *"You don't fall into a black hole. You fall out of phase."*

Disappearance is misalignment — not destruction.

❖ *"Stillness wins. Always."*

Because noise may shake the surface, but silence holds the structure.

❖ *"Ripple clarity is the new light."*

We don't need more speed. We need better alignment.

Observation and the Self

❖ *"Objectivity is a myth told by the blind to the blind."*

All observation is ripple-limited. There is no outside view.

❖ *"You didn't see the black hole — you saw where your mind gave up."*

Darkness isn't out there. It's within your threshold.

❖ *"Consciousness isn't inside you. You are inside a ripple field of awareness."*

You're not a container of mind — you're an expression of alignment.

❖ *"Death is not the end of life. It's the collapse of perception."*

The ripple continues. Only decoding fades.

On Science and Paradigm Shifts

❖ *"Math should not tell us what's real. It should follow what's resolvable."*

Equations are maps — not mandates.

❖ *"The singularity was never physical. It was intellectual surrender."*

When understanding hit zero, we called it infinite.

❖ *"Science mistook silence for absence and misalignment for collapse."*

The myth was born when certainty replaced curiosity.

❖ *"The greatest illusion in modern physics wasn't the black hole — it was the belief we understood it."*

Mystery became dogma too fast.

❖ *"We didn't escape darkness by adding more light. We escaped by removing distortion."*

The fix was never force — it was recalibration.

Living the Ripple Mind

❖ *"Don't force alignment. Find coherence."*

Whether in science or life, harmony isn't domination — it's resonance.

❖ *"When you stop resisting the field, you start living the truth."*

Truth is not fought — it is heard in silence.

❖ *"The ripple mind doesn't react. It resolves."*

It listens, aligns, and only then responds.

❖ *"You are not your thoughts. You are their alignment with the field."*

The self is not content — it is configuration.

❖ *"You don't find truth. You reduce distortion."*

Reality doesn't hide — we just stop seeing clearly.

Closing Note:

This quote bank is your field compass. Return to it when:

- The old narrative tempts you.
- The critics get loud.
- The silence gets deep.
- The next illusion presents itself.

Be still. Be sharp. Be ripple-aware. Because paradigm shifts don't begin with formulas — they begin with phrases that dare to reframe reality.

Appendix E: Einstein's 1939 Paper

Commentary and Reflection

"On a Stationary System with Spherical Symmetry Consisting of Many Gravitating Masses"

(*Albert Einstein, 1939, Annals of Mathematics*)

Historical Context

In 1939, Albert Einstein published a short but potent paper in which he directly challenged the then-developing notion of what we now call a black hole.

At the time, physicists like Oppenheimer and Snyder were exploring gravitational collapse scenarios — and laying the groundwork for the idea that massive stars could collapse into singularities.

Einstein wasn't convinced.

In this paper, he systematically constructed a model of a spherically symmetric cluster of matter and demonstrated that such configurations do not collapse into singularities under realistic physical assumptions.

His conclusions?

"The Schwarzschild singularity does not exist in physical reality."

He essentially argued that the conditions required for such a singularity are unphysical, and that nature prohibits such collapse.

Key Points from Einstein's Paper

1. **No Singularity in Reality:** Einstein made it clear that the Schwarzschild solution — which contains a singularity — should not be applied to describe the actual physical universe.
2. **Collapse Avoidance:** He concluded that gravitational collapse into a point of infinite density is theoretical only, and does not reflect any realistic system that can form in nature.
3. **Reluctance Toward Extremes:** He emphasized that the structure of space-time and the nature of mass-energy distributions do not allow such extreme behaviors without breaking physical laws.

UED Reflection: Einstein Was Right — But Overlooked the Field

In *Black Holes Don't Exist*, we honor this paper as the philosophical and scientific fork where Einstein resisted myth-making.

His intuition was powerful. His reluctance to accept collapse mirrored UED's core stance:

Nature does not collapse into nothing. It redistributes through ripple reconfiguration.

However, Einstein's framework lacked the Structured Vacuum Energy (SVE) model. He still operated within the geometrical fabric of General Relativity — a system based on curved spacetime rather than ripple tension.

Had Einstein seen the vacuum as a ripple-capable energy field, he might have translated his objection into a full-fledged alternative theory — much like UED.

Why This Paper Matters Now

Despite Einstein's stature, his 1939 dismissal of black holes was largely ignored or bypassed as the field moved toward theoretical glamor and mathematical idealism.

His clear warnings were overshadowed by:

- The seduction of singularities
- The growth of relativistic abstraction
- The success of observational proxies (like X-ray emissions near galactic centers)

But today — as paradoxes pile up and models hit walls — his hesitation looks prophetic.

Einstein knew:

The problem was not the mass — it was the math.

And now, with UED, we complete his resistance with a ripple-based model that retains structure, coherence, and realism without collapsing into theoretical blackouts.

Final Reflection: A Vision Interrupted

Einstein was not afraid to oppose his own framework when he sensed it was being misused.

His 1939 paper was not a denial of science — It was a defense of physical meaning over mathematical extremism.

In *Black Holes Don't Exist*, we carry that same torch forward.

We do not reject Einstein. We redeem him — by restoring his doubts and expanding his caution into a complete ripple framework.

Einstein sensed the problem. UED solves it. And the black hole dissolves — not with force, but with clarity.

Appendix F: Further Reading and References

Foundational Works Supporting the UED Paradigm

Author's Key Research Contributions

These reports and theoretical papers form the backbone of Universal Energy Dynamics (UED) and directly challenge mainstream interpretations of black hole physics, wave theory, and gravitational phenomena.

1. Chavan's Law: A Universal Energy Equilibrium Model for Black Holes Zenodo, 2025-02-01

DOI: 10.5281/ZENODO.14782900[1]

1. Formation of Black Holes: A Universal Energy Dynamics (UED) Perspective Zenodo, 2025-02-02

DOI: 10.5281/ZENODO.14788539[2]

1. https://doi.org/10.5281/ZENODO.14782900

2. https://doi.org/10.5281/ZENODO.14788539

1. Universal Energy Dynamics (UED): Empirical Research Phase 1 – Experimental Pathways and Data Reinterpretation Zenodo, 2025-04-29

DOI: 10.5281/ZENODO.15300391[3]

1. Universal Energy Dynamics vs Established Theories: A Comparative Framework for the Future of Physics Zenodo, 2025-05-15

DOI: 10.5281/ZENODO.15423130[4]

1. Chavan's Cosmic Continuum (C^3): Manifestation and Dissolution of the Universe within Structured Vacuum Energy (SVE) Zenodo, 2025-04-26

DOI: 10.5281/ZENODO.15287388[5]

1. Chavan's Structured Energy Equilibrium Principle (SEEP) Zenodo, 2025-02-24

DOI: 10.5281/ZENODO.14916491[6]

1. New Theory of Light: Ripple Alignment and the ATT Principle Zenodo, 2025-05-10

DOI: 10.5281/ZENODO.15377693[7]

3. https://doi.org/10.5281/ZENODO.15300391

4. https://doi.org/10.5281/ZENODO.15423130

5. https://doi.org/10.5281/ZENODO.15287388

6. https://doi.org/10.5281/ZENODO.14916491

1. The Myth of Wave-Particle Duality: A Structured Vacuum Energy Perspective Zenodo, 2025-02-03

DOI: 10.5281/ZENODO.14793162[8]

1. Universal Energy Dynamics: The Great Realignment Zenodo, 2025-05-03

DOI: 10.5281/ZENODO.15331312[9]

1. Chavan-Einstein Unified Theory (CEUT) Zenodo, 2025-03-08

DOI: 10.5281/ZENODO.14992289[10]

Recommended Foundational Readings for Context and Contrast

These classical sources are recommended for comparison, critique, and context — particularly in understanding the historical development of the black hole concept and the philosophical turn taken by modern physics.

- Albert Einstein (1939) *On a Stationary System with Spherical Symmetry Consisting of Many Gravitating Masses Annals of Mathematics, 40(4), 922-936*
- Roger Penrose (1965) *Gravitational Collapse and Space-Time Singularities Physical Review Letters,*

7. https://doi.org/10.5281/ZENODO.15377693

8. https://doi.org/10.5281/ZENODO.14793162

9. https://doi.org/10.5281/ZENODO.15331312

10. https://doi.org/10.5281/ZENODO.14992289

14(3)

- Stephen Hawking (1974) *Black Hole Explosions? Nature, 248, 30-31*
- Kip Thorne (1994) *Black Holes and Time Warps: Einstein's Outrageous Legacy* W. W. Norton & Company
- Sean Carroll (2010) *From Eternity to Here: The Quest for the Ultimate Theory of Time* Dutton Books

Author Profile & Access

Full list of works: Sandeep J. Chavan ORCID https://orcid.org/0009-0005-4190-7164

Author Website: https://sandeepjchavan.com

Final Note

This book is only one surface ripple in a much deeper field. The journey doesn't stop here. The questions deepen. The models clarify. The future awaits alignment — not just in theory, but in every thought we dare to revise.

Closing Note

The End of Collapse — The Return to Coherence

We began with a question — and ended with a revelation: That what we thought was the most terrifying force in the cosmos was simply a **misunderstood silence.**

The black hole was never a monster. It was a **mirror**, showing us the limits of our language, the distortion of our models, and the illusions we built from our own unresolved questions.

This book does not ask you to believe. It invites you to **re-see.**

It dares you to look beyond the drama of collapsing stars and into the stillness that always was — the **Structured Vacuum Energy (SVE)** field beneath all motion, meaning, and mind.

It offers not a new belief, but a **new balance**. Not a final answer, but a field in which **questions evolve through alignment**, not abstraction.

Universal Energy Dynamics (UED) is not a rebellion. It is a return — to coherence, to observation, to simplicity, to structure.

We close this book not at the end of science, but at the **beginning of a new scientific vision** — One where the ripple replaces the rupture, Clarity replaces collapse, And silence is no longer feared — but understood.

There are no black holes. Only boundaries waiting to be resolved.

Thank you for walking to the edge with me. Now let's turn back — not in retreat, but in realization. The universe was never broken. It was always whole. Now, so are we.

— Sandeep J. Chavan

Acknowledgements

To the Ripples Behind the Silence

Every book is a constellation of influences, challenges, and invisible hands — and this one, in particular, wouldn't exist without a field of alignment far beyond the self.

To those who doubted — thank you. Your resistance became the force that shaped my precision. This book is not a rebellion against you, but a resolution of the very questions you couldn't ask aloud.

To the thinkers who dared to pause where others rushed ahead — Einstein, Bohm, and others who whispered "something doesn't feel right" — this book is a tribute to your hesitation, your humility, and your unfinished thoughts.

To my students and readers — thank you for challenging me with your curiosity, your questions, and your stubborn need to make sense of the universe. You kept me honest.

To my family — your quiet understanding, patience, and unconditional support anchored me when the work grew loud. This silence was possible only because of your presence.

To every researcher who chooses clarity over consensus, coherence over complexity — may this book remind you: You're not alone in seeing differently.

To the ripple field itself — the silent, structured intelligence that flows beneath all form, force, and fiction —thank you for letting me decode a piece of your symmetry.

And finally, to the reader holding this book: Thank you for suspending belief, if only for a moment. If something in these pages shifted your perception — even slightly —then the ripple has done its work.

With stillness and sincerity,

Sandeep J. Chavan

About the Author

Sandeep J. Chavan is an independent thinker, educator, and researcher whose work challenges some of the most entrenched ideas in modern science. With a foundation in physics and a lifelong curiosity about the unseen architecture of reality, he has authored over **100 books** and **multiple research reports** across cosmology, consciousness, philosophy, and ripple-based field theory.

Sandeep is the originator of **Universal Energy Dynamics (UED)** — a bold new framework that redefines force, gravity, time, and even black holes as ripple-based interactions within **Structured Vacuum Energy (SVE)**. His work bridges science and philosophy with clarity, cutting through abstraction to return to structural truth.

He is known for **questioning the unquestioned**, for seeing **stillness where others see collapse**, and for restoring **meaning where modern physics offers paradox**.

His growing body of research, including the **Chavan-Einstein Unified Theory (CEUT)** and the **Ripple Alignment Model**, is hosted openly for global peer engagement. Through his writing, teaching, and public commentary, he invites readers to rethink not only what the universe is — but how we have been trained to perceive it.

Sandeep lives and works in India, where he continues to write, teach, and explore the boundaries of science, mind, and metaphysics — often all at once.

Learn more: sandeepjchavan.com[1]

Research profile: ORCID 0009-0005-4190-7164[2]

1. https://www.sandeepjchavan.com
2. https://orcid.org/0009-0005-4190-7164

Don't miss out!

Visit the website below and you can sign up to receive emails whenever Sandeep Chavan publishes a new book. There's no charge and no obligation.

https://books2read.com/r/B-A-EPGPC-QBSRG

BOOKS 2 READ

Connecting independent readers to independent writers.

Did you love *Black Hole Don't Exist*? Then you should read *Chavan's Theory of Everything*[3] by Sandeep Chavan!

[4]

Chavan's Theory of Everything: A Ripple-Based Redefinition of Reality

What if the universe is not expanding, not filled with particles, and not governed by forces as we know them?

What if motion is an illusion, time is a perceptual artifact, and energy is not a thing—but a consequence?

Welcome to *Chavan's Theory of Everything*, a paradigm-shifting scientific revolution that dares to rethink the very foundation of physics, cosmology, and consciousness.

3. https://books2read.com/u/bzNZxz

4. https://books2read.com/u/bzNZxz

Rooted in the **Universal Energy Dynamics (UED)** framework, this groundbreaking work replaces conventional assumptions with a simple, elegant idea: the universe is absolutely still—and everything we observe is the result of ripple disturbances resolving within a structured vacuum.

Written by **Sandeep Chavan**, an independent physicist, educator, and deep systems thinker, this book unifies scientific rigor with philosophical clarity. With over a decade of teaching experience and a research portfolio that spans cosmic mechanics, particle paradoxes, and consciousness studies, Chavan offers a model that dismantles quantum confusion, redefines gravity, explains light without motion, and resolves the long-standing contradictions between relativity and observation.

In this theory, there are no photons darting through space, no mysterious particles popping in and out of reality. There is only **ripple tension, alignment thresholds**, and the unfolding of **perceptual consequence**. From this ripple field arises energy, time, memory, and even the observer.

This is not just a theory—it is a lens. A tool to re-examine everything we once accepted as fact.

Inside, you'll explore:

The Law of Ripple ConsequenceStructured Vacuum Energy (SVE)Chavan's Limit and Alignment Threshold Time (ATT)Why energy is not created or destroyed—it is revealedHow perception, not measurement, defines realityThe bridge between ancient wisdom and cutting-edge science

Whether you're a scientist, student, philosopher, or skeptic, this book invites you to see again—with clarity.

You've been told the story of particles and spacetime. Now, step into the silence beneath it all.

Because the universe never moved.You did.

Read more at https://www.sandeepjchavan.com/.

Also by Sandeep Chavan

Chavan's UED research

CMGED Research Report

Chavan's Cosmic Continuum (C^3)

Frequency

Ripples of Reality

Reality Realigned

The Great Realignment

I – The Cosmic Leverage

Chavan's Cosmic Theory of Relativity (CTR)

Chavan's Theory of Light

Feeling Like a Physicist

TIME – The New Definition

$E \neq mc^2$: The Collapse of Classical Assumptions

The Chavan Codex

Chavan's Atomic Model

Chavan's Theory of Everything

Echoes of War: The Russia-Ukraine Conflict and Its Global Ripples

Shattered Lives
The War Economy
Borders Redrawn
Arms and Allies
Collateral Damage
Justice in the Ashes

Somewhere in the Darkness of the Cosmos
The Blueprint of the Universe
The Cosmic Breath
From the Edge of the Universe

The Eternal Sage Chronicles
The Sage and the Sovereign
The Family Pillar
The People's Flame
The Warrior's Soul
The Bridging Mentor
The Guiding Light

The Exodus
The Last Cycle: 2075

The Fundamental Theory of Life & Existence

Energy in Action: The Soul Demystified

The Active Mind

Energy in Connection

Death: The Disconnection

Subconsciousness: The Chosen Disconnection

Desires: The Engine of Life

The Restoration of Human Knowledge

Restoring Bhagavad Gita

The War of Thoughts

Cracks in the Known

The Human Game

The Divine Dilemma

The Original Intent

The Answers

Ultra Modern Life Philosophy

The Duality of Laughter and Tears

Take It Easy

Compare to Accept

Standalone

It's Not AI, It's AHI - Amplified Human Intelligence

The Decision Paradox
The Four Sapiens
Malicious Script of Indian Polity
The IIT Legacy & Global Impact
The IIT Legacy & Global Impact
You are just a Version of Your Original
Win the Game You Didn't Choose
Always Keep Your Bags Packed
Reveal to Shield - Learn the Game of Facades
Engage Beyond Elections
Culture, Identity & Change: The Evolution of Indian Society
The Triangular Dynamics
Never Trust Without Doubt
If You Can't Change the Country, Leave the Country
Live Fearless with Fear
India: The Investment Magnet
Realignment
Powering a Billion Dreams
Universal Skill
Extension: From Brush to Algorithm
Human Side of Einstein
The Future Knows No Lines
Let's Bring the Generations to the Table
Unified Energy Dynamics
A Beat That Never Dies
Strategic Horizons
Right From Scratch
The Forgotten Innovators
Physics Fixing
The Savior Illusion
The Selfish Side

The Age of the Weakest Sapiens
Neural Wars
The Absorption: There Is No Death
Strange Therapist
Bubbles: Become a Drop
Beyond the Blue
Half-Scholar
The Law of Truth
Failure
Philosophy 2.0: A Structural Manifesto
Chaos: A Gentle Guide to the Unresolvable
Self-Help: New Syllabus
Rippling
Game: How to Play the Share Market
Out of Syllabus
Saffronized: The Code Behind the Crown
The Anger
Moksha: The Liberation
Chavan vs Einstein: The Universe Was Never Curved
Rocket Science: The Ripple Gambling
The Cosmos: The Still Universe
The Taste of Silence
Chavan's Post-Materialist Human Model
10-Second Rule
Chavan vs Hawking
It was Never a Wave Signal
Quantum Courtroom: The Trial of a Century
Chavan–Heisenberg Ripple Uncertainty Principle
The Great Disruption
Dark ≠ Hidden
Breath

Calculus of Life
The Earth Doesn't Move
Black Hole Don't Exist
Chavan vs Kant

Watch for more at https://www.sandeepjchavan.com/.

www.ingramcontent.com/pod-product-compliance
Lightning Source LLC
LaVergne TN
LVHW090321210926
848480LV00001B/325
* 9 7 9 8 2 3 1 9 6 0 7 3 6 *